Canada's Arctic Sovereignty

Resources, Climate and Conflict

Jennifer Parks

CANADIAN
CURRENTS

© 2010 by Lone Pine Publishing
First printed in 2010   10 9 8 7 6 5 4 3 2 1
Printed in China

All rights reserved. No part of this work covered by the copy rights hereon may be reproduced or used in any form or by any means—graphic, electronic or mechanical—without the prior written permission of the publish ers, except for reviewers, who may quote brief passages. Any request for photocopying, recording, taping or storage on information retrieval systems of any part of this work shall be directed in writing to the publisher.

Lone Pine Publishing
10145 – 81 Avenue
Edmonton, Alberta T6E 1W9

Website: www.lonepinepublishing.com

Library and Archives Canada Cataloguing in Publication

Parks, Jennifer, 1975
      Canada's Arctic sovereignty / Jennifer Parks.

Softcover ISBN 978-1-926736-00-6
Hardcover ISBN 978-1-926736-03-7

      1. Arctic regions--Strategic aspects. 2. Canada--Boundaries--Arctic regions. 3. Canada, Northern--Economic policy. 4. Petroleum reserves--Canada, Northern. 5. Environmental policy--Canada, Northern. 6. Canada--Military policy. I. Title.

FC191.P37 2009   971.9   C2009-906005-1

Editorial Director: Nancy Foulds
Project Editor: Wendy Pirk
Production Manager: Gene Longson
Designer: Rob Tao
Layout: Rob Tao, Kamila Kwiatkowska
Cartography: Kamila Kwiatkowska
Photo Research: Randy Kennedy
Cover Design: Gerry Dotto
Cover Image: front, photos.com/Jupiter Images/Radius Images; back, Natural Resources Canada
Photo Credits: Creative Commons, Attribution Share Alike: United States Air Force/Senior Airman Joshua Strang 22; Digital Vision 78; Dreamstime 95; Yvonne Harris 226; photos.com/Jupiter Images 5, 10, 21, 27, 47, 82, 96, 109, 129, 137, 138, 168, 169, 196, 197, 201, 228

We acknowledge the financial support of the Government of Canada through the Book Publishing Industry Development Program (BPIDP) for our publishing activities.

Dedication

I believe that every right implies a responsibility; every opportunity an obligation; every possession a duty.

—John. D. Rockefeller, Jr.

This book is dedicated to the most sovereign authority of all—Mother Nature—and to the indigenous people of the north, who have lived and prospered by Her hand for millennia. May we learn to temper our capitalist ambitions with true knowledge of abundance...before She really gets cross angry with us!

PC: P13

Contents

Acknowledgements 4

Foreword 5

Introduction 10

Chapter 1:
Our True North Strong and Free—or Not? 21

Chapter 2:
Staking Claims: Whose North? 47

Chapter 3:
Global Warming: In Our Lifetime or the Day
   After Tomorrow? 82

Chapter 4:
The Race for Tomorrow's Energy:
   Untapped Oil North of 60 109

Chapter 5:
Defending the Far North 137

Chapter 6:
A Future for Northern Peoples 168

Chapter 7:
Rolling with Climate Change: Greener Industry,
   Sovereign Responsibility 196

Conclusion 228

Selected References 237

Index 240

Acknowledgements

This project would not have been possible without the help and support of many people. Thank you to Ken Davis and Nancy Foulds of Lone Pine Publishing, and to my editor, Wendy Pirk. I loved the process of bringing this book into being; it has stretched me as a writer, researcher and thinker. To Dr. Rob Huebert, for his time and insights on Arctic security and Canadian foreign and defence policy. To Ed Struzik, Randy Boswell, Doug Saunders, Ken Coates, P. Whitney Lackenbauer, William Morrison and Greg Poetzer, for their indepth and intelligent coverage of northern issues. To Lew and Jane Fahner, for all of their clipped articles and enthusiasm for my project. To dear Mr. Marm, who encouraged me to write—no matter what came out—and to Professor Michael Bristol, who instructed me to write with humility and discipline and "Murder your darlings" before they get in the way. A special thanks to my parents, John and Lessia Parks, who bought me a thesaurus when I asked for one, and taught me that powerful words are meaningless without the right action. To Kev, my soulmate in this wild adventure called life, and to Rod, Suz, Bry and Meg, who've always had my back. Your love is my real inspiration.

Foreword

The Arctic is changing in so many ways that it is very easy for most Canadians to simply give up trying to understand what is happening. We are constantly being bombarded by new and often contradictory information. "There is a race for resources in the Arctic"..."There is no race"..."The polar bears are vanishing"..."The polar bears are fine"...and the list goes on. The Arctic is central to our very existence as Canada. Our history, security, prosperity and even our understanding of who we are, are all tied to the Arctic. Since it is in a process of transformation, it is of absolute necessity that all Canadians understand what is happening both within and to the Canadian Arctic.

Jennifer Parks has written this book to explain to Canadians what is happening in their Arctic. Her book is not written in the language of academics, scientists or bureaucrats. Instead it is written in language that is accessible to all Canadians. This book explains how many forces in the Arctic are now coming together so that all Canadians can understand what is happening. It is an unsettling account because the story is still unfolding. We do not know what the ending to this story will be. It could have a happy ending or one that is not quite so happy. It depends on a lot of factors. But the one thing that she makes clear is that our leaders need to have a realistic understanding of what is happening. We cannot be blinded by wishful thinking when it comes to the Arctic. Ms. Parks makes it clear that the Arctic is too important.

Several important themes resonate throughout this book. First and perhaps most importantly, the Arctic is important to Canada because Canadians live there. What happens in the Arctic affects the security and well-being of those who call it home. For some Canadians, the Arctic has been their home for thousands of years. Thus, to talk about Canadian sovereignty and security as if it was only some abstract notion does a disservice to all Canadians. If there is a threat to Canadian Arctic sovereignty, there is a threat to Canadians. For example, both the Americans and the Europeans claim that the Northwest Passage is an international strait. This claim is rightly known as one of the most serious challenges to Canadian sovereignty. But why does it matter? It matters because if the Americans and the Europeans have their way, then Canada loses the ability to control international shipping in its Arctic waters. It is not simply a dispute over legal definitions. If Canada loses, then we lose the ability to prevent ships we perceive as dangerous to the waters from entering. We also lose the ability to create Marine Protected Zones, which would protect a specific region where we may want to stop all shipping from entering.

Canada does not want to totally prevent international shipping in the Northwest Passage. It wants the ability to control the manner by which this shipping proceeds. Canadians, in particular northern indigenous Canadians, understand the region in a way that few others do. Southern Canadians are often guilty of forgetting this important point. But if northerners are to be properly consulted (and not merely patronized) to ensure that Canada's northern waters are properly protected and used, Canada needs to retain control over the region.

Control of the Northwest Passage is not the only issue that Canada faces. We also have northern maritime border disputes with both the United States and Denmark. The outcome of these disputes will determine what can and cannot be done in the disputed zones. As we move to determine the outer limits of our continental shelf, it is possible that we may face overlaps with Denmark, the United States and Russia. Of course, we do not yet know what our claim or that of our neighbours will be. We may come to a shared and complimentary agreement as to where the boundaries need to be drawn, or there could be serious disagreements over overlaps of our claims.

Jennifer Parks also makes it clear that Canada must be prepared to meet the new actors that are arriving in the Arctic. The combination of melting ice cover and the promise of great resource wealth means that many foreign actors will want to seek their fortunes in the Arctic. The promise of great oil and gas wealth is especially tempting in a world of continually increasing demand. While it may be that the promise of this wealth may outstrip the reality, it is clear that many non-Arctic states and actors are now preparing their plans. South Korea has emerged as the world's leading builder of ice-capable commercial vessels. Japan is investing heavily in developing means of exploiting the gas hydrates (a potentially huge source of new energy resources) that are found in northern waters. China is expanding its polar scientific capabilities and is sending its large polar research vessels into Arctic waters off Canada's northern coast. The European Union is talking of the need to create an international treaty that would oversee the control of the resources of the Arctic Ocean. Some of these developments are good for Canada; others are not. The reality is that Canada needs to fully understand what each and every one of these actions means for Canadian interests. Where it favours Canadian interests, Canada needs to welcome and embrace the new arrivals. But where the actions of outsiders harms the interests of Canada and northern Canadians in particular, then we must be firm in rebuffing their efforts at intervention.

However, for Canada to be prepared, our government needs to give the north its full attention. Both the Martin and the Harper governments have indeed given the Arctic a high priority. Nevertheless, past Canadian governments have become aware of the problems facing the north and made a series of bold and intelligent policies, only to allow them to fade away as other issues arise. Jennifer Parks has made it clear that Canada no longer has the luxury of ignoring the north and hoping that its challenges will go away. One hopes that the Harper government will not turn away from its current course of action in the Arctic. However, the only way to ensure a steady course is to ensure that the Canadian public understands the magnitude of the challenges and opportunities that now await Canada. An educated and understanding public is the best guarantee that should the government decide to claw back on its current action, there will be sufficient public outcry to bring them back on course. At the same time, an informed Canadian public can also provide the necessary support for the more difficult and expensive decisions that must be made. Thus this book performs an important service by discussing a complex subject and putting it in terms that can be understood by all Canadians.

—Dr Robert Huebert

Introduction

It's an unseasonably warm day in June of 2009. I can hear the din of traffic in the streets below as I talk on the phone with Dr. Rob Huebert, currently the most sought-after Canadian expert on Arctic issues. It has taken weeks to get this interview.

As we finally chat, the faint smell of exhaust rises from the hot tarmac, wafting through my office window. I push the glass closed and crank the fan. This kind of sweltering afternoon in Edmonton prompts long-time advocates of global warming to nod their heads knowingly, then speed-dial their brokers in Toronto to buy up more energy security stocks.

For a guy who's been intensely busy for the last four years—leading conferences, serving on advisory boards, participating in top-level working groups and giving countless on-the-fly media interviews—Huebert sounds surprisingly relaxed, even refreshed, as we speak.

Perhaps it's because I've reached the father of three at his home in Calgary, where I can hear his twins and six-year-old playing nearby in the background. He obligingly discusses at length his views on Arctic security, circumpolar relations, Canadian defence policy and the nation's claims to resource riches in the north—stopping only once to heed a tiny, talkative voice by the receiver.

Seconds later, Huebert is diving passionately into a confab of what he sees as the pivotal issues facing a warming world that is running on empty and looking north to exploit its virgin stocks.

The climate change debate is not just about melting ice and endangered polar bears, he emphasizes. It concerns a global problem in urgent need of a global solution.

The polar race is heating up as anthropogenic—or human-induced—climate change boosts temperatures, melts Arctic glaciers and opens a previously hostile region to new maritime travel, tourism and a frenzy of oil, gas and mineral exploration.

An estimated quarter of the world's last petroleum and gas deposits lie beneath waning Arctic ice. Nothing short of a prohibition on human ingenuity, and greed, will stop the advances of industry into this area—not even extreme and unpredictable weather conditions, such as the erratic storms, thinning ice and melting permafrost that are signalling the rapid procession of climate change.

Nations are scrambling for a piece of the "black gold" bonanza in the High North to feed the global addiction to oil. The increasing stress placed on already fragile eco-systems is destroying natural habitats, endangering wildlife and threatening northern communities and their traditional way of life.

Meanwhile, the Arctic is becoming a hotbed of mounting military interest, which will have many local and global consequences in the coming years, the defence expert tells me.

Huebert is the associate director of the Centre for Military and Strategic Studies (CMSS) and associate professor in the Department of Political Science at the University of Calgary. He is a strategist, a political thinker, a writer, a teacher, an environmentalist and a tireless advocate for safety and security in the Canadian North and around the world.

During our conversation, he draws a surprising but apt parallel to a classic game-theory parable—the Prisoner's Dilemma—that he learned when he was a university student.

The Prisoner's Dilemma is a behavioural metaphor based on tit-for-tat logic. Psychologists, political scientists, biologists and economists use it as a problem-solving model for analyzing and understanding the dynamics of competition and—more importantly—co-operation.

"If you don't prepare and your opponent does, you stand to lose the most. But if you trust your opponent to co-operate and take the collective interest into account, it can be a win-win situation. The question is, can you trust your opponent to do that?"

He is, of course, referring to the circumpolar drama unfolding in our daily news headlines. Does the Arctic belong to Canada? Should the U.S. have maritime access through the Canadian North? Is Russia a trustworthy neighbour—or is an old Cold War bear waking up from two decades of hibernation? When the ice finally breaks up, who will get the biggest shares of oil and gas up north, and with it, untold petro-power? Will energy-hungry China and India get a seat at the table?

Although there's no imminent threat of northern conflict, says Huebert, some nations are cautiously preparing for the worst by beefing up homeland security and surveillance systems, diversifying domestic energy portfolios and establishing national emergency fuel funds.

No one wants to be first to hit bottom-barrel, especially in the event of a full-blown energy war. But "militarizing" the situation could drive up tensions and make it difficult to bridge political distrust or build consensus on issues, he says.

Meanwhile, the race is on to develop clean technologies that will power our future. As we reach a phenomenon called "peak oil"—the cresting and declining of world oil supplies—yesterday's oil barons are investing their petrodollars in tomorrow's green industry. Forward thinkers are heralding the Age of Oil as nothing but a blip in the course of human history.

A blip that, right now, fuels our world.

While this prophecy may turn out to be true, today's reality tells a different story.

As long as there is thick, black oil in the earth, and as long as we can build the high-tech tools required to extract it, a nation's very identity in the eyes of the world will remain closely tied to its ability to generate wealth and compete in a global market fuelled by oil.

Climate change is making the Arctic more accessible to outsiders. Consequently, Canada must increasingly defend its northern limits and justify its claims to coveted energy resources in the circumpolar north, says Huebert.

Back when the country's "true north, strong and free" was lodged in thick, multi-year ice, no one bothered challenging Canada's authority over its Arctic region. No one cared. It was too cold and cut off. Pristine and vast—but limited in value. The weather is changing this attitude.

While Canada already has quite a long history of mineral extraction in the Far North, it's only been over the past 30 years that we've seen a major surge in the development of diamond mines, oil and gas and other mining-related activity. Today, there is a rich and soon-to-be-accessible cache of oil and gas in the Arctic, and other nations want a piece of it. As a result, Canada's character as a northern nation and its corresponding sovereignty rights in that region are being thrown under the microscope. The view, under this magnification, is not all that flattering.

Scientists predict the Arctic could soon be ice-free for the first time in history, unlocking significant amounts of natural resources and energy wealth. Other nations are suddenly scrutinizing Canada and its custodial track record up north. They are using its spotty record of defending its northern interests over the years to challenge the nation's jurisdiction in parts of the region and gain legal access to its waterways and potential wealth.

Huebert thinks Canada has a lot of work to do to secure its role and rights in the Arctic. The country's on-again, off-again commitment to the north, its history of ignoring real northern issues and its financial inability to respond on its own to

Introduction

Canadian Arctic security threats have, in the eyes of others, tarnished Canada's image as a truly northern nation, he says.

The unapologetic realist blames Canada's southbound gaze and its over-reliance on the U.S. to defend its northern border—first from the Japanese in the '40s, then from the Soviets following World War II—for the current challenges being raised over its maritime rights in the Canadian North.

"Canada has always tried to do things in the north on the cheap because the Canadian northern vision is coast to coast—not coast to coast to coast; it lies within a 200-mile (320-kilometre) radius of the American border and in the capitalist economy," Huebert argues.

In his many writings on the issue, he discusses the circumstances driving what he calls a "renaissance" in Canada's Arctic security.

Climate change and its numerous impacts—the rush to exploit oil and gas reserves in the Canadian North, the contemporary threat of terrorism to North American security and the high profile that national media is now giving to sovereignty and security issues—are issues that are starting to converge, and none will go away any time soon, says Huebert.

This dynamic crossroads we're at is propelling Canadian policy-makers to take a new look at northern security and redefine Canada's relationship to the north.

Huebert goes on to explain that it is Canada's duty, right now, to prove itself as an Arctic nation by rising to the challenges rapidly reshaping the north as we know it.

In this book, you will discover what those challenges are. You will read about the Arctic as it was, how the face of the new Arctic is rapidly changing, and what forces are driving this transformation.

Many Canadians take for granted that the north belongs to Canada; that the Arctic's idyllic icebergs, midnight suns and northerners swaddled in fur-lined parkas are as intrinsically Canuck as our red-and-white Maple Leaf Flag.

We Canadians have a dilemma: other countries with competitive interests in the Far North are raising the stakes. They are levying challenges against not only Canada but also against one another in a play for northern power and coveted non-renewable energy resources.

Russia already has a long history of Arctic exploration and industrial activity in the region and has audaciously staked out the North Pole. Denmark argues that the polar pinnacle, among other disputed regions, is Danish, while Norway has secured a large swath of sea territory and is looking to legally extend its Exclusive Economic Zone. So are Russia and Denmark. Canada and the U.S. are collaborating on their own scientific research to do the same—despite other major differences.

Meanwhile, the U.S. and European Union (EU) are pressing for unrestricted access to the Northwest Passage, a key waterway in the Canadian North. They argue that global safety and security are at stake, and that Canada should set aside its national interests in favour of the larger picture.

Even non-Arctic nations, such as China and India, are trying to buy their way into the action.

Against this frenetic backdrop, the rapid advances of global warming are slowly but surely transforming the free and fabled Arctic from a once-glistening, relatively undisturbed landscape into an ecological disaster zone. The industrial boom in the area, which began 30 years ago, is wreaking havoc on the traditional lands and lifestyle of Canada's largest Inuit population. This trend is growing. Northerners risk losing the most from current and future resource exploration in their region.

A once-remote and forbiddingly cold Arctic is, today, a blazing beacon on everyone's radar. The dust under Canada's proverbial northern rug is getting exposed. At present, Canada's sovereignty as a northern nation is anything but secure.

As the ice melts and oil prices soar, a chain of reactions will put Canada's integrity as an Arctic nation to the test.

This image is not the idyllic portrait of Canada's strong, free and unsullied north that was painted for us in grade-school textbooks. What remains to be seen is if Canada will step up, and whether other nations will pursue their self-interests or join together to deal with pressing matters.

Finally, I ask Huebert what he thinks the Arctic will look like in another 30 years. He pauses a moment to reflect, then replies:

> It will be a much busier area....With the expansion of resource development, we will see the transformation of the indigenous

society....The north won't necessarily be the focal point of conflict but will increasingly be an area of spillover tension. The biggest trouble will come from the unexpected political consequences of actions by people not thought to be that important at all.

Counter to what his critics may think, Huebert prides himself on being an optimist, but times like these require a wise man's reserves of pragmatism.

He concludes:

If we're lucky, in 30 years the Arctic will be a golden example of how to co-operate and do things properly. I hope we will have developed a governance system that works. I hope we will have made the Arctic secure—not just in terms of its military safety but also in terms of the environment, culture and people of the region.

This web of issues is woven through the northern security debate that's going on today. Throughout the pages of this book, you will "listen in" on the fascinating, colourful conversation taking place between politicians, scientists, policymakers, environmentalists, lawyers, human rights advocates, business leaders and everyday people.

You will learn how this dynamic dialogue has sparked nationalist rows and over-zealous acts of patriotism, and how it is paving the path for new partnerships and strategic innovations that could make our world healthier, safer, more efficient and sustainable.

You will meet some formerly complacent leaders who are being inspired into decisive action and see

Introduction

why this action could change how Arctic nations define global security issues and protect their circumpolar borders in the near future.

You will be challenged to weigh "scientific" evidence against grand hyperbole, and wade through the often murky waters of legal litigation and bureaucratic spin-doctoring.

Splashy news headlines and public agendas won't tell the whole story. You will be guided to read between the lines and consider the rest of the story; it is still being written in the public eye, behind closed doors, out in the field, on the roads, in our mega-stores, in our homes and in our heads.

Whether you study politics or economics, prefer science fiction thrillers, Rolex watches, hour-long showers or walks in nature, this conversation concerns you. You are part of it. We all are.

Its outcome will affect our energy bills, our health, the cost of our California fruit and whether the "exotic," after all, comes at too high a price.

It will affect the weather, the seasons, our sense of safety and our sense of ourselves— Canadian, northerner or otherwise. And it will most definitely irrevocably alter our relationship to the earth, the sun, the wind, water—and oil.

In the coming months and years, likely sooner rather than later, we will be forced to learn which sources of sustenance we can—and can't—live without. Meanwhile, Canada's role and rights in the north are being called into question.

How our nation's leaders act must be unambiguous and committed to securing its northern borders and interests, says Huebert. He adds that, especially in these reckless and uncertain times, the outcome will be flavoured, no doubt, by what's going on around us.

"We must ensure that the Arctic region benefits and is protected for all Canadians, but northerners in particular. Canada must have the ability to control what happens in the area of what it calls the Canadian Arctic."

In these pages, you will find out what you need to know to not just follow, but also to join, the ongoing conversation that's emanating from the Arctic as it unfolds. Since it's part of your story, why not help write the ending?

CHAPTER ONE

Our True North Strong and Free—or Not?

The Arctic—it's a place unlike any other. The vast, rugged beauty of its glacial landscapes, its exotic wildlife and its harsh, unforgiving climate have captured the imaginations of Canadians for centuries. Since early explorers risked their lives to chart the far, fabled trade routes of the polar region, our northern vision has glistened an icy cobalt blue under the lidless skies of a distant world, unspoiled by us. Few Canadians call it their "home and native land," but the Far North—with its iconic icebergs and polar bears, Inuit *inuksuit* and aurora borealis—is a cherished part of our cultural heritage. We regard the region, its inhabitants and its natural wonders with a benign sense of patriotism and protectiveness, as a rare and final frontier, relatively untouched by the momentum of modern life and industry. The Arctic—our "true north, strong and free"—has been a steadfast symbol of Canada's national identity to its citizens and most of the world—until now.

Humans are rapidly altering the natural environment. We're on the cusp of an ecological shift that is transforming life on the planet and reinventing our social, cultural and geopolitical realities.

The frigid Arctic has moved from the margins of global awareness and now sits at centre stage, a political hot potato in an international debate on sovereignty: who is the rightful owner of the land, water and people of our northern reaches?

The unfolding story—well underway—is a high-stakes ecological drama, a 21st-century odyssey of humans, industry and nature. The conflict is set in the present, where the future balance of planet Earth rests on the integrity of a global politics that calls profit "progress," often at the expense of the environment. The mercury is rising....

Climate change is melting the Arctic, and what was once considered a stunning but useless stretch of ice north of Canada's mainland—running from Alaska to Greenland through the Arctic Archipelago islands—is now a beacon on everyone's radar.

Some scientists predict the Northwest Passage, which provides a circumpolar shortcut from North America and Europe to Asia, could be open for year-round shipping as early as 2040, or for seasonal trips in a decade or less.

A slumping worldwide economy and stiffer market competition make the so-called "Arctic Grail" a coveted trade route for commercial mariners seeking efficient bottom lines in tougher times.

By travelling through the Northwest Passage instead of the Panama Canal, shippers can cut 7000 kilometres off of a trip from New Jersey to Shanghai, saving both time and money.

The early sea-faring explorers who had their visions set due north were onto something; the untapped wealth and opportunity in the Arctic are enormous—if you can only beat the cold and ice to get there.

Historically, the famed route was impassable because of long winters and a solid ice pack up to 8 metres thick—too dense, dangerous and costly to penetrate with icebreakers and drill ships.

But climate change is opening up the Northwest Passage. As the hard ice melts, what's left is softer and easier to break up, making navigation in warmer months a likelihood in the near future.

Meteorological data shows that higher average temperatures worldwide are changing the face of the Arctic we know. According to the National Snow and Ice Data Center (NSIDC), summer sea ice in the Arctic Ocean has shrunk at an alarming rate, from an average of nearly 8 million square kilometres between 1979 and 2000 to a little less than 4 million square kilometres. These statistics point to one conclusion: we're on the fast track to a totally ice-free Arctic.

The warming that began with the build-up of greenhouse gases in the earth's atmosphere—primarily because of the burning of human-made fossil fuels and deforestation—is having a much more pronounced effect up north, say scientists.

The top of the world is heating up, and ice is melting faster because of dramatic regional changes to the earth's albedo, the fraction of the sun's energy that is reflected back into space.

A quick look at the basic science of ice-albedo feedback helps explain the speed-up of Arctic thaw.

Sea ice is white. Ocean water is much darker. Sea ice reflects most of the sun's radiation, whereas the dark ocean water absorbs most of the sun's energy. When water heats up, the ice melts faster. As ice melts and exposes more ocean, the planet heats up at an accelerated rate. Scientists say Arctic warming is now driving the pace of climate change around the world.

Nearly a decade ago, the United Nations' Intergovernmental Panel on Climate Change (IPCC), a multinational scientific body supported by the United Nations, predicted that the Arctic could be ice-free by 2100. Now they believe it could happen as early as 2070.

For the first time in recorded history, the Northwest Passage was "fully navigable" during brief periods of summer, the European Space Agency reported in September of 2007.

Climate change is melting the Arctic, but, ironically, the industrious spirit that's tipped nature's scales is solid and intact—and eagerly finding opportunity amid crisis. Once the deep ice broke, it didn't take long for the great vessels of commerce to start launching their ships, heralding a "new era in Arctic shipping."

In the early dawn of October 18, 2007, the *Kapitan Sviridov* docked at Canada's northern seaport of Churchill, Manitoba, carrying a shipment of fertilizer bought by a Saskatchewan-based farming co-operative. The cargo came from the seaport of Murmansk in Estonia, Russia, and was the first successful shipment across the "Arctic bridge"— a long-discussed marine route between Canada and Russia that wasn't navigable a few years ago.

In a *Globe & Mail* report of the historic crossing, Canadian Institute of International Affairs researcher Michael Berk remarked that global warming is creating an enormous opportunity for trade and transport along what is potentially the shortest sea corridor between North America and Eurasia.

"If we expand and connect Churchill and Murmansk, an ice-free, year-round port, we're talking about creating a bridge that will link North American markets with increasingly important Eurasian markets," said Berk.

A shorter oceanic delivery system will make trade partners that were less than desirable in the past seem much more appealing. Already, voyages through the Northwest Passage are starting to redefine the boundaries of international maritime trade and travel, sway political agendas and peak national interests.

When the *Kapitan Sviridov* departed the Churchill port, it was carrying another shipment: Canadian wheat destined for sale in Italy. Recognizing the port's strategic value, Prime Minister Stephen Harper has committed $68 million to upgrading Churchill and the railways linking it to the prairies. The future will likely see more activity and growth in that region of Manitoba.

In 2008, the MV *Camilla Desgagnes* was the first commercial cargo to travel through the Northwest Passage, delivering supplies from Montreal to communities in western Nunavut.

The Nunavut coast guard had an icebreaker on standby, in case of any problems, but the inaugural journey by transport company Desgagnes Transarctik was incident free, a CBC news report stated.

"I didn't see one cube of ice," said Waguih Rayes, the company's Arctic Division general manager, who was aboard the vessel. The conditions would have been enviable to the early die-hard explorers who battled blizzards and gales—many to their death—to navigate the icy northern passage.

The Canadian Ice Service—a branch of the Meteorological Service of Canada (MSC)—contends that the Northwest Passage is not likely to become a reliable shipping route for decades because of extreme ice variability. But the motor is rolling, and other commercial transporters are expected to follow in the MV *Camilla Desgagnes*' wake. Some experts think seasonal traffic in the maritime passage will surge as early as the next decade.

The tourism industry has also jumped on board, offering scenic cruises off the coast of Greenland, Alaska, Russia and Nunavut and through the long-fabled Northwest Passage. The breathtaking views they're selling are so priceless—and costly—because they are framed by a grave, poignant paradox. What travellers witness is both untouched beauty and endangered landscape; rugged and wild, yet vulnerable to the human forces of exploitation; a glimpse of the eternal—but only for a limited time.

Today's Arctic, as the average Canadian knows it, romantically, is an vague reality that tourists,

scientists, adventurers, artists and nature lovers are scrambling to capture for posterity.

No ice, after all, means open water, and open water means unfettered access to the untapped natural riches lying beneath the Arctic seabed.

The world is now looking due north to Canada's "true north," with visions of crude barrels and dollar signs dancing in their heads.

"This opens up the possibility of enormous oil and gas resources, larger than those stored in the Middle East, available to whoever can claim to own the roof of the world," wrote *Globe & Mail* columnist Doug Saunders, who has covered polar politics extensively.

He wrote, in the October 20, 2007, article, "Treading on Thin Ice":

> It's a dark irony that the burning of fossil fuels that contributed to the ice melt may end up making it possible to extract another century's worth of Earth-warming petroleum reserves from beneath that ice. But it's a possibility that is driving countries like Canada to spend millions.

An international debate is heating up as governments jostle to stake their claims on the Arctic, and the oil and minerals beneath it. As the ice cover shrinks, some controversial questions are emerging.

Who owns the Arctic? Is she Canada's "home and native land"? What claims do we have to her waterways and resources under existing international law? Does the Northwest Passage belong to Canada or to the world?

Furthermore, will Canada's Arctic sovereignty transcend climate change and circumpolar conquest—or will it melt away with the last cerulean blue ice floe?

Canada regards the Northwest Passage as "internal waters" and claims the sovereign right to assert control over all activity within those waters.

Other countries—namely the United States, Russia and nations of the European Union—reject this claim outright. They view the Northwest Passage as an "international strait" that is open to all nations for free, unrestricted transit.

Canada's loudest opponent, the U.S. argues that any water between two open seas must be open to all shipping. The Northwest Passage links the Atlantic and Pacific Oceans and, as the ice melts, it will offer a more direct and sheltered shipping route around the globe.

"We don't recognize Canada's claims to those waters," said U.S. Ambassador David Wilkins at a political forum in 2006. At the time, he was opposing Prime Minister–designate Stephen Harper's plan to boost military presence in the Arctic, one link in the current chain of initiatives by the Conservative government to assert Canada's Arctic sovereignty before it's too late. Wilkins supported Harper's move for more security, but he opposed the Canadian position on the legal status of the Passage.

Historically, the U.S. has shown blatant disregard for Canada's claims to the Arctic waterway. Despite high costs and extremely icy conditions, the Humble Oil icebreaker *U.S. Manhattan* travelled through the Northwest Passage in 1968 and 1970 without seeking Canada's permission.

Then in 1985, the American icebreaker *Polar Sea* coasted through the icy waters without so much as asking its northerly neighbour first (not very neighbourly), resulting in a heated diplomatic dispute. They did ask to plan the trip; they did not ask permission to make it.

To drive the debate home, in 1987 during then-President Ronald Reagan's visit to Ottawa, then-Prime Minister Brian Mulroney reportedly pulled out a globe in his office, put his finger on the Northwest Passage and bluntly told the president, "Ron, that's ours." He showed the president the ice that covered that portion of the globe, pointing to its unique nature.

Following the voyage of the *Polar Sea*, the two countries signed the Arctic Cooperation Agreement, stating that the U.S. agrees not to send any more icebreakers through the Northwest Passage without prior Canadian consent, and that Canada will always be willing to grant its consent.

The icy waterway at the top of the world first enticed explorers with its fabled shortcut hundreds of years ago.

In the mid-16th century, English voyager Martin Frobisher sailed three times in search of the Northwest Passage as a trade route to India and China, spurring Britain's fur trade and colonialism in Canada. He never found it. Many others made attempts, including Sir John Franklin, whose lost expedition in 1845 remained shrouded in mystery for over a century. It wasn't until 1906 that Norwegian Roald Amundsen became the first explorer to sail the entire length of the Northwest Passage. The glory was great, the Grail, in the end, a bit disappointing. The famed route—no longer fabled—turned out to be too icy, dangerous and costly for travel.

The strategic significance of the Northwest Passage then waned until the 1950s, when the mounting threat to Arctic security by Communist Soviet Union saw the building of the Distant Early Warning Line, or DEW Line, a radar system that was established to alert the military of approaching Soviet bombers and subsequently missiles during the Cold War. In 1957, the year the DEW Line construction was complete, the American vessel *Storis* became the first U.S. ship to circumnavigate North America via the Northwest Passage.

Today, it's not the thrill of conquest, or even the famed eastern shortcut that's creating a buzz, but rather what lies beneath the water.

Five Arctic-basin nations—Canada, the U.S., Russia, Denmark and Norway—are vying for one-quarter of the world's untapped mineral and energy reserves, which are known to lie beneath the receding Arctic ice.

There are an estimated 90 billion barrels of oil and nearly 3 billion cubic metres of natural gas to be found in the Arctic, much of it in offshore areas, according to the U.S. Geological Survey (USGS).

International law will decide who can claim it. Under the United Nations Convention on the Law of the Sea (UNCLOS)—a treaty that defines the perimeters of nations' governance and use of the world's oceans—a nation can legally control sea territory up to 12 nautical miles (22 kilometres) from its coastline. (A nation still needs to allow for innocent passage even in territorial seas. It is only internal waters that state has complete control of [i.e., sovereighnty] including that of shipping.)

A nation's official coastline is determined by straight baselines—lines drawn from key coastal points to delineate the nation's maritime boundaries. The baseline system was introduced after an international court ruling in 1951—involving Norway—established that the 12-nautical-mile (22-kilometre) limit could be extended, in some cases. One such instance is if a nation's coast is dotted by islands and has "internal waters" spanning greater than 12 nautical miles (22 kilometres) from shore to shore. Canada has made such a case to protect its sea territory between the deep coastal indentations of its mainland and throughout the archipelago islands.

At least 150 nations have ratified UNCLOS, created in response to nations that wanted to extend their coastal boundaries and make legal claims to natural resources, protect wildlife and enforce pollution regulations within those boundaries.

By early 2009, the U.S. was the only major country that had not ratified the convention, except landlocked Switzerland and a few other countries such as Israel. But growing American interest in the Arctic's economic jewels is creating speculation that President Barack Obama will ratify the agreement soon so that the U.S. can officially enter its claims. Canada ratified UNCLOS in 2003. (We signed it in 1982 and were a major force in its creation.)

In 1986, following the *Polar Sea* incident, Canada established straight baselines around its Arctic Archipelago, which include the waters of the Northwest Passage.

However, this action did not clear up the confusion over the passage's rightful owner; it's still regarded today as an international strait by some nations. Several of the archipelago waterways span up to 100 kilometres wide—an attractive legal loophole for a foreign ship to cruise through.

Experts predict that as the ice melts and industry stakes rise, the passage will become more susceptible to clandestine voyages.

The Law of the Sea treaty states that a nation has exclusive control of the resources under its coastal waters for up to 200 nautical miles (370 kilometres) from shore and an additional 150 nautical miles for the control of the soil and subsoil of the seabed if it can show it has a continental shelf. The treaty also states that if a country wants to extend its territorial waters, it must prove that a geological area—in this case the Arctic seabed—is an extension of its continental shelf. The country may then claim the resources lying beneath it.

These days, it's a royal tea party underwater for the nations of the so-called Arctic-5, each of which is busy gathering seabed data to build their sovereignty case, while intermittently politicking and attempting to keep abreast of each other's activities. The ubiquitous presence of nuclear submarines in the Arctic Ocean and a head-to-head race for the best sub sonar–detection equipment in the deep sea are enough to raise red flags and set off Cold War alarm bells along the DEW Line. But these missions are "officially" in the name of science, not national defence, say government officials.

On such a research trip, in August of 2007, the U.S. sent the 420-foot icebreaker *Healy* into the Arctic to map the sea floor north of Alaska and determine the extent of its continental shelf. Coincidentally—or not—Russia made a controversial move one week before the U.S. ship embarked on its data-gathering voyage: it sent a submarine down 4000 metres through the Arctic ice and planted a Russian flag on the North Pole seabed.

Why make such a bold statement? Russia claims it has scientific proof that it owns the Lomonosov Ridge, a mountain chain running across the Arctic Ocean floor and directly beneath the North Pole, spanning from Siberia to Ellesmere Island and Greenland, north of Canada.

The U.S. dismissed the Russian move as legally worthless, whether it stuck "a metal flag, a rubber flag or a bedsheet in the Arctic floor," media reports said. Canadian Foreign Minister Peter Mackay called the deep-sea voyage "just a show," and mused over whether Russia really thought it could claim territory using "15th-century" tactics.

On his way to an Arctic-5 meeting in Ilulissat, Greenland, on May 27, 2008, Russia's foreign minister Sergei Lavrov compared the move to Neil Armstrong planting a U.S. flag on the moon in 1969. "There is no claim for any territory," he reportedly said. "There couldn't be because there is the Law of the Sea Convention and there are mechanisms created to implement this convention."

Moscow already submitted its claim once, in 2001. The UN Commission on the Limits of the Continental Shelf indicated that it needed more data.

Russia says it owns 45 percent of the Arctic. Norway claims to have majority stakes in the Barents Sea. Canada is eyeing rich reserves within the Arctic Archipelago. Denmark says it has geological proof that its Nordic land extends all the way to the North Pole—and claims the polar pinnacle as its own.

Meanwhile, the U.S. and Canada are disputing Alaska-Yukon maritime boundaries and oil and gas rights in the Beaufort Sea. Massive stakes in the Arctic have even led Canadians and Danes to set aside differences over who owns Hans Island—a tiny yet controversial island between Canada's Ellesmere and Denmark's Greenland. They are collaborating on their own geological study of Lomonosov Ridge, which could make both nations wealthier. After gathering research, the two countries announced in 2008 that they have strong evidence the ridge is an extension of North America.

The five Arctic nations have 10 years from the date they signed the UNCLOS treaty to submit scientific data supporting their claims to an extended continental shelf, for review by the UN Commission. Russia and Norway's deadline is 2009, Canada has until 2013 and Denmark has until 2014.

No one owns the North Pole—yet. The top of the world is still considered high seas—an international region open to all who can get there.

In the end, the country—or countries—that can prove its continental shelf is connected to the Arctic seabed will gain economic access to the earth's last supply of untapped non-renewable resources and will hold untold political power as the energy crisis of the future unfolds.

The wheels of industry are cranking into full gear in India and China, and experts say that as the supply of oil and gas starts to decline, petroleum prices will spike with rising demand until pipes run empty or alternative energy sources are brought to the market.

Some believe this circumpolar race to carve up the Arctic seabed may, in the end, be a tie—seeing five nations dividing the region's industrial wealth and sharing the burdens of inevitable environmental debt.

Currently, there are two scenarios on the table for the division of land, water, resources and responsibility:

1) The "median line method," which would divide up the Arctic between countries based on the proximity of their coastlines to the North Pole. Denmark would gain the North Pole but Canada could benefit considerably from this scenario also.

2) The "sector method," which would divide the Arctic using north-south lines drawn from the North Pole. This option is not as desirable for Canada, but Norway and Russia stand to profit from this scenario.

Today's energy race won't hinge on conquistador-style seizure of land and water—unlike the days of colonial expansion—but rather on "geoproximity," what may be the political buzzword of this century.

The unique geography at the top of the world is affecting the dynamics of current Arctic politics and giving new meaning to another present-day turn of phrase—"global village."

If you look down on the North Pole from space, it becomes clear that countries conventionally thought to be on opposite sides of the world from one another—Canada, Russia, Norway, Denmark and the United States—are actually circumpolar neighbours. These neighbours just so happen to share one giant backyard, which just so happens to be full of buried treasure!

When the ice melts, the "digging" season will begin, and every nation on the block wants the green flag to get in there.

Our unique Arctic-5 village—its sovereignty bids, deep-ocean mapping, political alliances and legal wrangling—are making the determination of coastal rights in the polar region both complex and fascinating, says Peter Croker, chairman of the UN's Commission on the Limits of the Continental Shelf, an international body set up to arbitrate the process.

"It's the only place [in the world] where a number of countries encircle an enclosed ocean. There is a lot of overlap. It you take a normal coastal state, the issues are limited to adjoining states and an outer boundary. In the Arctic, it is quite different," he told the British Broadcasting Corporation (BBC) in 2005.

"Geoproximity" will play a vital role in defining the Arctic's future—it's people, environment and the divvying up of resources. Although the word and concept didn't exist in 16th-century European vernacular, it was in fact Canada's irresistible "geography" and stone's-throw "proximity" to rich natural resources—fur, fish, forests, minerals and waterways—that led Europe to explore and conquer its lands and waters.

A brief look at Canada's colonial past reveals how a future nation got its sovereignty. The imperial beginnings of our young nation also provide much-needed context for understanding a debate that's challenging a number of Canada's Arctic sovereignty claims today.

Canada—called New France— was a possession of France from the mid-1500s, when Jacques Cartier sailed to the shores of "New Found Land" and planted a holy cross bearing the French Coat of Arms. In 1763, Canada was traded in a peace deal brokered by Great Britain, France and Spain. The Treaty of Paris ended the Seven Years' War and also a century's struggle between France and Great Britain for control over our vast resource-rich lands. The war's conclusion marked the beginning of British rule outside of Europe. The Hudson's Bay Company, which was created in 1670, continued its jurisdiction over all land draining into Hudson Bay.

Over the decades, Britain spread its power around by developing strategic trading posts and communities across Canada. Expansion forced natives off their lands and onto "reservations," a tactical move that laid the basis for today's current reserve system.

The central pillar behind Britain's sovereignty over Canadian territory was built on the colonial belief that native peoples had no legal claim to the land they lived on, only the right to live off the land and what it produced. This stultifying notion has yet to be reconciled for First Nations and Inuit peoples in today's courts.

The next century brought changes, new colonists and conflict that helped Canada on down the path toward independence. There was the American Revolution, the creation of English-speaking Upper Canada and French-speaking Lower Canada, the War of 1812 and an emerging middle class of immigrant colonists who pressed for their own constitutional rights and self-government. Several rebellions and riots later, the British parliament passed The British North America Act in 1867 and created the Dominion of Canada. Canada still had colonial status under the British Crown, but it was now "officially" it's own country.

The Hudson's Bay Company's land—including present-day northern Quebec, Manitoba, much of Ontario, Saskatchewan, Alberta, the Northwest Territories and Nunavut—was sold to the new Dominion in 1870, giving Canada full control over of its northern lands—all but the Arctic Islands, that is.

Here's where the dispute over Canada's Arctic sovereignty begins. No one doubts the status of Canada's mainland. But it's a number of the thousands of Arctic Archipelago islands located north of the mainland that are questionably Canadian. Some of them don't even officially belong to us!

A number of the islands were first visited by British, American or Scandinavian explorers. When Britain ceded its land holdings to Canada—including the archipelago islands discovered by Brits—it also passed along islands that were uncharted or had been explored by other countries.

The Canadian Encyclopedia online states:

> In July 1880, the British government transferred to Canada the rest of its possessions in the Arctic, including 'all Islands adjacent to any such Territories' whether discovered or not—a feeble basis for a claim of sovereignty, since the British had a dubious right to give Canada islands [that] had not yet been discovered, or [that] had been discovered by foreigners.

The concept of a nation's sovereignty is rooted in international law. Sovereignty, according to the UN Charter, is founded on the following main pillars: owning and controlling or administrating a territory; use and occupancy of the land; and exercising responsibilities in that area with regard to justice, commerce and, increasingly, human rights and environmental protection.

A famous quote by Lassa Oppenheim, a leading authority on international law, sums up why the debate over Arctic sovereignty, like many others, is so complex and hard to resolve.

> There exists perhaps no conception the meaning of which is more controversial than that of sovereignty. It is an indisputable fact that this

conception, from the moment when it was introduced into political science until the present day, has never had a meaning [that] was universally agreed upon.

The debate over Canada's sovereignty in the Far North is further complicated by criticism—both at home and abroad—that the nation has historically shown only token interest in the Arctic region and has, at times, even completely ignored its issues and responsibilities in the north.

This criticism includes the following points:

After the controversial U.S. *Polar Sea* voyage, External Affairs Minister Joe Clark tabled plans for a brand new $500 million icebreaker, but it was never built. The 1987 White Paper on Defence announced that Canada would buy 10 to 12 nuclear-powered submarines and "polar class 8" icebreakers to beef up underwater surveillance in the Arctic, but the costly plan was scrapped at the end of the Cold War. A 1996 program to boost Canada's underwater presence in the Arctic was later abandoned as a result of cost-cutting. In December 2005, the Canadian media reported that a U.S. nuclear submarine passed through Canadian territorial waters while touring the Arctic Ocean, "possibly without permission from the Canadian government." Critics argued that if Canada were more assertive in the north, the U.S. would not be challenging Canadian control of the Northwest Passage.

Other nations are using such criticism to call into question Canada's sovereignty claims in the Arctic while bolstering their own legal cases.

International law states if a sufficient number of transits are made without Canada's permission to qualify the passage as a "useful route for international marine traffic," the Northwest Passage could be deemed an international strait, overriding Canada's internal waters claim.

This understanding would seriously restrict Canada's ability to monitor shipping as well as environmental and defence issues along its maritime borders.

A parliamentary report in 2006 recognized the need to assert and protect Canada's northern sovereignty with swift actions, namely increasing deep-water surveillance in the region and putting in place policies to control who can travel through the Northwest Passage.

The report concludes:

> Some have argued that, given the challenges posed by monitoring such a vast territory, Canada's resources are insufficient in terms of ensuring the capability to enforce its sovereignty in the region....Future policy discussions will need to consider the most effective and efficient means of protecting Canadian sovereignty in the Arctic, including assessment of what could be potentially costly programs.

Canada's willingness to acknowledge its shaky sovereignty status in the Arctic has shifted over the past decade, says Dr. Rob Huebert. In his 2001 article, "Climate Change and Canadian Sovereignty in the Northwest Passage," Huebert argues that Canada has focused disproportionately on climate change as the driving issue in Arctic politics today.

He contends that the government has ignored the main issues—growing international doubt over Canada's claims to Arctic sovereignty, and the urgent need to assert our presence in the north (or, as the Stephen Harper government slogan now pledges, "use it or lose it").

In his article, Huebert cites a speech by Canada's Legal Affairs Bureau official Mark Gaillard, which was delivered in Whitehorse on March 19, 2001. He writes:

> He [Gaillard] argued that Canadian sovereignty over the waterways of the Canadian Arctic did not depend on the ice cover of the region, but that Canada's view, then and now, is 'that since the 1880 deed transfer [of the Arctic Archipelago from the U.K. to Canada], the waters of the Arctic Archipelago have been Canada's internal waters by virtue of historical title. These waters have been used by Inuit, now of Canada, since time immemorial. Canada has unqualified and uninterrupted sovereignty over the waters.'

The tone of his speech is unanimous and unwavering—just what the spin doctor likely ordered—but it takes for granted that Canada's Arctic sovereignty is a *fait accompli*.

However, it's not a done deal. Not everyone thinks Canada owns the Arctic. In fact, the legal challenges are numerous. Huebert notes that Gaillard failed to acknowledge this reality. Today, both historic titles and collective cultural sentiment are being challenged by cold, hard legal fact. In the end, the UN will decide whose claims stand in the True North. Our Arctic, today—melting or not—may not be our Arctic tomorrow.

On March 11, 2009, also in Whitehorse, Canada's Minister of Foreign Affairs, Lawrence Cannon, delivered quite a different speech to the world concerning our Arctic foreign policy:

> My government has invested significantly... to ensure that Canada secures recognition for the maximum extent of its continental shelf...Canada is an Arctic nation and an Arctic power. Canada's Arctic and north make up over 40 percent of our landmass. We occupy a major portion of the Arctic. The Arctic and the north are part of our national identity...

The tone of this speech is frank and clear but concedes much—namely, that Canada must prove its entitlement to parts of the north; that it is not the unanimous landlord of the Arctic but is a cool-headed shareholder among five circumpolar nations, just looking to maximize and protect its future stakes in the region.

Cannon details the many multi-million dollar initiatives that the Harper government has been rolling out since 2006, when it pledged on its party platform to defend Canada's Arctic sovereignty by affirming "leadership, stewardship and ownership in the region."

These initiatives include building a "world-class research centre" and a new army training centre and refurbishing an existing deep-water port, all in the Arctic; initiating a plan to improve living conditions for northern First Nations and Inuit people; finish mapping the Arctic continental shelf;

amending the 1970 Arctic Waters Pollution Prevention Act to extend the zone for regulating shipping activity and practices in Canadian territorial sea; and creating new regulations under the Canada Shipping Act to require "mandatory reporting" for all vessels planning to voyage through Canadian Arctic waters—namely the Northwest Passage.

Will Canada's delayed response to the warming debate on Arctic sovereignty be enough to protect our national assets, or will we lose out to other countries with greater strategic ambitions in the north?

Law of the Sea treaty expert Rosemary Rayfuse, a professor at Australia's University of New South Wales, gave her take on Canada's odds in the sovereignty debate in the *Globe & Mail*:

> As a matter of legal principle, Canada has no greater claim on the Arctic than any of the other four Arctic states, or indeed than any state in the world, quite frankly.
>
> Depending on geological realities, Canada may have a geological claim on some extension of the continental shelf, although I think that Russia and Norway have the greatest amount, from what I've seen.

Canada has several pending territorial disputes with the U.S., one with Russia and two or possibly three with Denmark, all of which may seriously hamper Canada's ability to preserve its current claims to the region.

In Rob Huebert's pre-released chapter of the book *Northern Exposure: Peoples, Powers and*

Prospects for Canada's North, the Canadian Arctic sovereignty expert says that Canada will lose control over those claims if it does not act decisively to resolve standing issues.

> While most Canadians believe that our national claim is incontestable, the reality is that Canada is embroiled in numerous international disputes over various aspects of control over its Arctic region....Should some or all of them have unfavourable results for Canada, the international community may come to view Canadian claims of sovereignty with skepticism.

Global warming is creating a wealth of new economic opportunity north of 60, and the battle for the Arctic is heating up. As the ice melts, it is opening access to a coveted polar trade route and the world's last cache of energy and mineral resources. The Arctic-5 are on their marks and ready to stake their claims. The Law of the Sea will serve as the compass for the circumpolar debate. The UN treaty commission, in the end, will deliver the sovereignty verdict.

Meanwhile, several long-standing and emerging issues—environmental, socio-economic, geopolitical and cultural—are converging to vault the Arctic to the top of the world's political agendas.

Says Huebert: "A perfect storm is brewing, and it's prompting the Canadian government to act."

What will we do, as a nation, to "stand on guard" for our Arctic sovereignty?

CHAPTER TWO

Staking Claims: Whose North?

A snapshot of the Arctic today is quite unlike the frozen pictures of the past.

The "Big Melt" or "Great Thaw" of our age is well underway—both in the north and around the globe. Lavishly snow-capped peaks, popular with tourists, are retreating in the Andes, Himalayas and on Mt. Kilimanjaro. Peru's Cordillera Blanca ("White Mountain range") may soon need to change its name.

A look at the big picture reveals a world in transition: international trade, power politics, national security, border and seabed disputes, indigenous issues and environmental change are all dynamic factors in a warming global debate over rights to land and water.

Skip ahead half a century and take another look. The top of the world, seen by space satellite, is blue, not white. The long-melting polar ice cap has disappeared entirely, and nature's original skyscrapers—the extraordinary, jagged azure green icebergs that once speared a remote, forgotten landscape—exist only as surreal images in your grandchildren's natural history textbooks.

Alaskan snow crabs are off the menu at Red Lobster. In fact, they no longer exist, but for top dollar you can splurge on Russian snow crabs. These tasty crustaceans, which depend on nutrients from algae growing beneath ice that sinks to the ocean floor, left Alaska and went northwest toward Russia when the polar ice vanished from American waters.

Here, in this new Arctic frontier, the sun sets on a changed skyline. Offshore oil towers are the neo skyscrapers. The cold, wild yonder has been colonized by industry. Polar bear "jails," located in several of the thriving oil and gas towns, capture and anesthetize those that wander into urban areas—forsaken national icons fallen to "intruder" status. The drug wars have moved farther north, with narcotics traffickers reaching indigenous communities via international Arctic shipping lanes, in which the Royal Canadian Mounted Police have limited jurisdiction. Money—clean or dirty—rules this northern frontier.

Two emergent power nations—Norway and Russia—are the biggest energy distributors in the world, and fuelling up costs triple what it used to. Most of the time, though, you choose to ride your bike to avoid rebuke from your "green" grandkids, who think of your generation as a bunch of environmental bullies with politics for brains and piggy banks for vision. On the bright side, average temperatures in February have risen by 10° C, and tougher emissions standards mean you can breathe as you peddle past traffic jams at rush hour.

This scenario, while imagined, is entirely possible. Today—right now—temperatures are rising, the ice is melting, animal populations are shifting, political interests are hungrily eyeing northern resources, critical borders are ill-defined and native communities are increasingly vulnerable to imminent exploitation and change.

The current race to carve up the ocean floor, however, is not limited to the Arctic. Any nation that belongs to the 1982 Law of the Sea Convention had until the end of 2009 to submit claims that clarify or officially extend its continental shelf.

In total, more than seven million underwater square kilometres are in question—roughly the size of Australia—ranging from the Arctic Ocean and South China Sea to Antarctica.

"This will probably be the last big shift in ownership of territory in the history of the earth," said Scandinavian scientist Lars Kullerud, an advisor to nations making submissions with the GRID-Arendal foundation, a non-profit body run by the UN Environmental Program and Norway.

Most of the claims controversies are with groups of countries that ring or border a body of water, like in the Arctic. For example, China, the Philippines, Vietnam, Malaysia and Taiwan all have competing claims on the Spratly Islands, situated in the South China Sea, which are, likewise, confounded by colonial histories and national defence ambitions.

How land and water get divvied up in the Arctic and beyond will set the course for the future. Impenetrable Arctic ice may have staved off the sovereignty debate for centuries, but climate change has brought it to the fore. The contenders are wielding yardsticks, historical records, national flags and nuclear subs to back their claims "campaigns." For the most part, they're doing so civilly, but zoning rules in this neighbourhood are far from settled, and as the *New York Times* put it so succinctly, "if the Arctic is no longer a frozen backyard, the fences matter."

As this debate unfolds, there are several territory disputes left to settle between Arctic-5 nations. They include the tiny, uninhabited Hans Island—both Canada and Denmark claim this bump of rock between Canada's Ellesmere Island and Denmark's Greenland; a strip of the Beaufort Sea—the U.S. and Canada are disputing over maritime boundaries and a "wedge" of oil-rich sea on the Alaska-Yukon border; the Barents Sea—Norway and Russia claim the same economic zone; the Bering Sea—Russia contends the U.S. has taken more than 50,000 square kilometres of resource-rich lands between Alaska and Siberia away from Russia; the North Pole—Denmark, Canada and Russia are trying to prove their respective continental shelves extend to the Lomonosov Ridge seabed beneath the North Pole, where vast reserves of oil, gas and minerals are known to exist; the Northwest Passage—or the "Canadian Northwest Passage," as Canada plans to rename it, to symbolically solidify its control over the disputed waterway.

Whose Hans?

Hans Island is a small, rocky knoll less than 100 metres wide that sits in the middle of Nares Strait, exactly halfway between Greenland and Ellesmere Island. Denmark and Canada have argued for more than three decades about who owns it. In 2005, the two nations appeared to have settled the matter, but since then both have behaved as though Hans Island is theirs.

In his exposé on the issue for *National Geographic*, Arctic expert Kenn Harper wrote:

"The island is barren and steep-sided. No one lives there. No one except scientific parties ever have. The question one is inclined to ask is not, 'Who owns it?' but rather, 'Who would want it?'"

The BBC has called Hans Island "a mouse that roared."

When Denmark and Canada drew a boundary line between their two territories in 1973, to officially divide up the underwater continental shelf between Ellesmere Island and Greenland, the line went down the centre of Nares Strait and Kennedy Channel. Canada tried to claim Hans Island as its own, but the Danes wouldn't have it. Finding themselves at an impasse, and not wanting to slow down negotiations, the two nations decided to leave the island out of the agreement.

The treaty line they drew didn't go around Hans Island or through it but simply skipped over it, stopping abruptly at the southern tip of Hans Island and starting again at the top, effectively leaving out 875 metres of border in the Nares Strait.

Undefined borders—especially when billions of barrels of oil potentially lie beneath them—aren't such a great idea. The problem of Hans Island didn't just go away.

By chance, in 1983, the Arctic expert who later wrote the exposé for *National Geographic* met a scientist in Resolute, Northwest Territories, who was wearing a hat with the embroidered words, "Hans Island, N.W.T."

These words puzzled Kenn Harper, who considered the island a part of Greenland, so he struck up a conversation with the scientist, who, he soon learned, worked for Dome Petroleum and had just returned from a summer of conducting ice research on Hans Island.

"Dome Petroleum, it turned out, had been doing research on this tiny island for some years. It planned to build offshore artificial islands on which to position drilling rigs in the Beaufort Sea, 1700 kilometres away. Hans Island was a surrogate for an artificial island," Harper later wrote.

Because Hans Island was the first obstacle that giant ice floes would hit when they entered the Kennedy Channel from the north in summertime, it was a good "test" to see how strong an artificial island needed to be to withstand bombardment by multi-year ice drifting down from the Arctic Ocean.

It was a brilliant plan, but when Denmark's Minister to Greenland caught wind of Dome Petroleum's research activity, he flew by helicopter to Hans Island and put up a Danish flag and left a bottle of schnapps and a note that read, "Welcome to the Danish Island."

The "battle of the bottles" began.

In July of 2005, Canadian soldiers paid a visit to the disputed island, raised a Canadian flag and left an Inuksuk—an Inuit stone sculpture—and a bottle of rye whiskey.

Were these gestures assertions of ownership or peace offerings—or a bit of both?

A week later, amid a flurry of rumours and rhetoric in the Canadian media that Denmark was stepping on national territory, Canada's National Defence Minister, Bill Graham, flew to the island during an Arctic tour of military stations. The Danish government was up in arms. It made a formal complaint to Canada for unlawful advances on its territory.

Denmark then sent the fishing patrol boat HDMS *Tulugaq* on a sovereignty trip to Hans Island, one of many trips made over the years to plant a new flag, replace a tattered one or leave signature Danish libations. The subtext of this trip seemed to be: "Keep your hands off Hans Island!" But, surprisingly, days later, foreign officials from Denmark, Canada and Greenland agreed to meet that September in New York to discuss matters during the UN General Assembly.

An editorial in 2005 by the Canadian American Strategic Review (CASR) called the Royal Danish Navy's occupation of Hans Island a test of "Canadian resolve."

However, Harper wrote in *National Geographic* that the diplomatic dispute made little sense at all:

Because it has already been agreed that the island will have no territorial sea, there is no possibility to extend one or the other nation's claims for offshore drilling or fishing rights. One suspects that Canada's unyielding position today has more to do with its claim to the Northwest Passage than to Hans Island itself. Canada appears to feel that losing its claim to Hans Island may set a precedent for challenge to the more important trans-oceanic passage through the heart of the High Arctic.

At the 2005 UN summit, Canada and Denmark announced they'd made an agreement to work together and resolve the issue.

Canadian Foreign Affairs Minister Pierre Pettigrew took the opportunity to restate that Hans Island belongs to Canada and to clear up any speculation that Hans is viewed by the government as just another symbol of Canadian sovereignty in the Arctic. Pettigrew said:

> It remains our firmly held position that Hans Island constitutes part of Canada's national territory. As long-standing allies, it is our shared objective to lead by example and resolve this matter, which we agree is about the island, and the island only.

Ironically, history reveals that neither Canadian nor Dane first occupied Hans Island. Canada has built its claim upon the basis of British discovery and its 19th-century inheritance of Britain's Arctic territories. But Hans Island wasn't discovered by a Brit. It was first found by an American named Charles Frances Hall, on August 29, 1871.

Northern Greenland belonged to the U.S., and Hall named the Island after Hans Hendrik, an Inuit from southern Greenland who was guiding his expedition.

It wasn't until 1916 that the Americans turned over their claims in Greenland as part of a deal involving $25 million and a transfer of the Danish West Indies—now the Virgin Islands—to the U.S.

The issue of Hans Island is unresolved. Both countries have since spent millions on geological studies to map the ocean floor beneath the disputed area, the now-famous speck in Nares Strait.

Beaufort's "Wedge"

The Beaufort Sea is another "grey zone" riddled by an unresolved boundary dispute, stretching back for decades.

This contentious body of water, which sits north of Alaska, the Yukon and the Northwest Territories, used to be frozen year-round. In this case, as with the others, global warming is cranking up the heat on an old drawing-room debate over national jurisdiction. As multi-year ice breaks up and summer navigation becomes quite feasible, the intellectual sparring over boundaries has suddenly become more real and consequential, with both sides scrambling to stake claims on underwater wealth.

In 2008, the Canadian Ice Service posted images of a giant fracture in the multi-year ice that covered the Beaufort Sea. Big chunks were starting to break off, move around and pull apart newer ice.

David Barber, a University of Manitoba climate specialist who was heading up an international research project on the Beaufort aboard the Canadian Coast Guard ship *Amundsen*, described his shock at the rapid melt of recent years to Canwest News Service (CNS).

The CNS article stated: "There is now so little thick, multi-year ice left that it is being blown around the Beaufort 'like Styrofoam in a bathtub,' says Barber."

Here's the dispute: Canada thinks the maritime boundary should follow the Alaska-Yukon treaty of

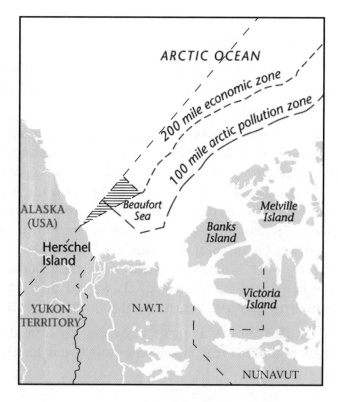

1825, which set the 141st meridian as the land and sea boundary between the two territories. A similar meridian line is used to divide U.S. and Russian waters to the northwest.

The U.S., on the other hand, believes that the boundary should be "equidistant" between their two shorelines. In question is a triangular-shaped wedge of water 30 degrees to the east of the 141st meridian with high energy potential. It has historically been used by Canada, but both countries have put up bids for petroleum exploration in the disputed area. Neither side is budging.

First, to secure its rights to offshore areas of interest, the United States must sign off on the Law of the Sea Convention, the UN pact that regulates all facets of maritime law, including commercial shipping and deep-sea excavation.

"While the treaty is supported by many U.S. groups, some conservatives don't like the idea of international law dictating American interests," wrote The Associated Press.

To move forward, the U.S. and Canada must both submit science-based claims for review by the UN Committee; where there are conflicting claims, a diplomatic deal will have to be struck. Diplomacy, however, may prove quite a challenge, given the high economic potential and the sovereignty issues at stake.

In August of 2008, Canwest News Service reported the Beaufort's northern waters were a veritable "hotbed of international activity."

That summer, Canadian Coast Guard icebreaker *Louis St-Laurent* was mapping the seabed, and Dr. Jacob Verhoef, the geoscientist heading up the expedition, described the Beaufort as "a very busy place."

Research icebreakers from China, Japan, Germany and the U.S. were out on similar data-gathering voyages. Everyone wanted to surmise the financial opportunity in the Beaufort region. The media played up the "security implications" of such unprecedented vessel activity in controversial waters. Canada's Defence Minister Peter MacKay shifted the debate from foreign ships posing a security threat to the need for effective patrolling and monitoring of such a widely desirable drilling area.

In July of 2007, Imperial Oil and ExxonMobil Canada bought 205,000 hectares on Canada's side of the Beaufort Sea for $585 million Canadian, the biggest investment to date in the region.

One year before, the U.S. independent Devon Energy Corp. had completed drilling at a Beaufort well, north of Inuvik. They reportedly spent $60 million Canadian, but the return on natural gas was "disappointing." Although Devon Energy didn't find the trillions of cubic feet of natural gas they'd hoped for, they did turn up 240 million barrels of recoverable oil.

Needless to say, ongoing interest in the Beaufort region will be a sure thing for years to come.

Rob Huebert, an expert on Arctic issues, bluntly told *Oilweek* in October of 2007 that the boundary dispute between Canada and the U.S. would not likely be resolved easily.

"I know Foreign Affairs is trying to pretend it's not a big issue, but what Canadian government is going to be willing to reach some form of a compromise agreement? What American government is going to be willing to give up oil and gas? That one's going to get really, really ugly," he said.

The 2008 U.S. Geological Survey estimated that 90 billion barrels of oil and 47 trillion cubic metres of natural gas remain to be discovered in the Arctic. In the Beaufort Sea, there may be as many as 30 billion barrels of oil and 6 billion cubic metres of natural gas waiting to be harvested.

If it weren't for the melting ice, officials might be tempted to ignore this stubborn issue a little longer, but clearly the stakes are too high for even a "wedge" of ambiguity.

A Barents Compromise?

Another decades-old dispute is brewing—this time in the Barents Sea, where Russia's and Norway's maritime borders meet. These northern waters were named after a Dutch explorer and later used by Cold War Russia as a ballistic missile bastion. The two nations have clashed since the mid-'70s over the Svalbard Archipelago—a jagged trio of islands by the names of Spitzbergen, Hopen and Bear Island.

This conflict is somewhat less diplomatic than the others, with a history of high sea chases, Soviet intelligence agents and reckless talk of war and trade embargoes.

Can tiny, determined Norway stand up to great Russian might? Some experts think this case may be decided, in the end, by the UN International Court of Justice in The Hague. Others believe diplomacy will prevail—but that power politics are inevitable.

The two countries claim overlapping economic zones within a 176,000 square-kilometre swath of water that's desirable, both for its renewable fishery resources and its large, lucrative fossil fuel potential.

Both nations have already made oil and gas discoveries in the Barents Sea. Snoehvit is the major $58-billion liquefied natural gas (LNG) plant on the Norwegian side, run by Norway's StatoilHydro since 2002, exporting to Europe and U.S. markets. Shtokman gas field, in the Russian sector, is being developed by Russian gas giant Gazprom and StatoilHydro with supplies starting in 2013. The Barents area holds great economic opportunity, but further developments are being forestalled by this long-standing boundary dispute.

Norway claims full control of the sea surrounding the Svalbard Archipelago. It is basing its argument largely on the Svalbard Treaty—also known as the Paris Treaty—which recognized Norway's sovereignty over the Svalbard Archipelago in 1920. This pact permitted other signatory nations the privilege of fishing and mining in Svalbard's surrounding waters, subject to Norwegian laws. The Svalbard Act of 1925 made the archipelago an official part of the Kingdom of Norway. In Norway's view, this act secured their sovereignty over Svalbard's land and water.

Their position was reinforced in 1982, when the Norwegian government claimed an exclusive 200-nautical-mile (370-kilometre) area for fishing and other economic development under UNCLOS. In Norway's view, this new treaty solidified the Paris pact and legally granted them unique rights to the zone while trumping the past privileges of other signatories. But several of these nations—including Russia—don't see it in such a light.

"While Norway claims that the Paris Treaty of 1920 gives it the right to establish an economic zone around the archipelago, Russia and a number of other countries maintain that the Paris Treaty only regulates the situation on the archipelago and not at sea," wrote the *Barents Observer* on May 8, 2008.

Russia and Norway have based their arguments on two opposing principles for defining boundaries: equidistant median lines and meridian—or sector—lines. Norway favours the former, which draws borders equally distant from each nation's coastal borders. Russia favours the latter, which follows international north-south meridian lines.

After World War II, in 1947, Soviet leader Joseph Stalin signed a treaty with Germany, which drew a line from the North Pole south to the Port of Murmansk, along the 35th meridian east, carving out a new Russian border. This line gave Germany more land—greatly welcome after its huge post-war territory losses to the Allies—and secured German support in the event that Russia ever went to war with Japan.

"Stalin once just drew a line from Murmansk to the North Pole and then to Chukchi and said,

'U.S.S.R. Polar Region'—and nobody worried about it," said Artur Chilingarov, an Arctic explorer and deputy for Russia's House of Parliament, in the *New York Times*.

The article, appearing in October of 2005, stated: "Now, instead of Stalin, the lines will be drawn by an international commission and the geography of the seabed itself."

The post-war Soviet border shift subsequently overlapped with Norway's territory, designated by median lines not recognized by the Russians. The resulting so-called "grey zone" of these two conflicting principles includes a disputed 155,000 square-kilometre area. Russia and Norway have agreed how to divide about 80 percent of it, but the rest is "no man's land."

In January of 1978, Norway's Minister of Trade and Shipping, Jens Evensen, signed a "grey zone" agreement—dubbed the "gold zone" by Soviet media—with the Soviets giving them an extended fishing season every year within the disputed waters.

"Many people in Norway believe that the grey zone agreement was a concession to the Russians and that Evensen signed the agreement under the pressure of his deputy Arne Treholt, who was unmasked as a Soviet intelligence agent in 1984," wrote maritime expert Pavel Prokhorov in a report for The Baltic Research Centre.

His remarks were presented at the 2006 Northern Research Forum (NRF)—a meeting of policy-makers, heads of state, researchers, community leaders and business people.

Prokhorov recapped the spike in tension between Russia and Norway in 2001, when the Norwegian Coast Guard detained the Russian fishing vessel *Chernigov* off the coast of Spitzbergen. The confrontation occurred within Norway's 200-mile (370-kilometre) "fishery protection zone," demarcated by Norway in 1977, without permission from the other Paris Treaty signatories. Russia acted quickly, sending in its warship *Severomorsk* to rescue the captive vessel.

Prokhorov stated in the NRF report:

The Norwegians did not want an armed conflict, and the *Chernigov* was released. Our etatists hailed the precedent and, when the Norwegian Coast Guard chased the trawler *Elektron* last year [2005], they called for not standing on ceremony and showing the Norwegians the might of Russian arms instead. Some even went as far as to say that, sooner or later, Russia would have to go to war against Norway for the Barents Sea.

Prokhorov concluded it would not be easy to win the case against Norway, but that Russia was left with little choice given that a small yet economically sovereign nation such as Norway could not be bullied by embargoes.

"There seems to be no reasonable alternative. After all, we cannot even stop gas deliveries to the Norwegians. They have gas of their own."

In 2007, Russia and Norway signed an agreement delimiting part of Norway's eastern tip, known as Varanger Fjord, which gave Russia an additional 26 kilometres of maritime territory, official reports stated. This zone, however, falls outside of the disputed area, so it may have been a peacemaking gesture.

It is unclear to what extent the two countries have reconciled their boundary dispute, but media reports suggest progress has been made.

"Head of state-owned oil major Rosneft last week indicated that Russia in the next 40 years will invest up to 12 trillion RUB in hydrocarbon exploration in disputed waters," wrote the *Barents Observer* on April 30, 2008.

"Only minor parts of the area are now believed to remain unsettled."

The Dow Jones Newswire announced on June 9, 2008, that upcoming talks between Russian and Norwegian foreign officials would "ratify an agreement reached last year which drew a 70-kilometre border around their immediate maritime zones."

During the two-day bilateral talks, the ministers reportedly discussed joint patrol and monitoring of the disputed waters for illegal fishing and transport and Russia's fishing rights within the Svalbard Treaty zone. Also on the agenda were the use of forbidden nets on Russian boats and Norway's concern over Russia's less-than-green pollution track record in the Arctic.

Although no official deal has been struck, the two nations seem to be inching closer to a settlement. Russia's deadline to submit its continental shelf claims to the UN committee is 2009. Both sides have huge stakes in the Barents Sea, overlapping in more ways than one—Norway's StatoilHydro holds a 24 percent stake in Russia's giant Shtokman gas field.

Norway may be small, but it's a powerful, progressive nation. Norwegians are the third-largest net exporter of oil in the world, behind Saudi Arabia and Russia. Althought Russia is big and strong, its clout in the Barents region substantially rests on the decisions of an international tribunal. Where there are conflicting continental shelf claims, Russia and Norway are on their own to resolve the matter.

The Barents Sea zone is believed to contain vast quantities of oil and gas, and neither nation wants a freeze on future exploration. Also, the rest of the world is watching, including nations such as the U.S., India and China, which are considering investments in Russia and Norway's waters; so a fine balance is required, taking into account future stakeholders.

All of this suggests a Barents compromise is not far off.

A Bering Sea Catch

Fish and oil are at the heart of a dispute between Russia and the U.S. in an area known as the Northeast Passage, which runs along the Russian coast and through the Bering and Chukchi seas.

The U.S. says Russia is claiming too much seabed and exclusive fishing rights in a zone owned by Americans since the 19th century.

In the mid-1990s, a village elder in a small Inuit outpost by the Chukchi Sea, close to the Russian border, called the U.S. Department of Fisheries and Wildlife to report "a massive fish kill," wrote the *New York Times*.

Everyone thought there had been a toxic spill at a nearby zinc mine, but it turned out it to be the end of a pink salmon run.

"They were dying of natural causes, as they always do once they spawn," a Fisheries and Wildlife officer reportedly said.

"The elders had never seen a run of this salmon species. But they have shown up every year since," chronicled the *New York Times*.

Global warming and ice melt are driving pink salmon and other fish and crab stocks that are normally found in American waters around Alaska farther northwest toward Russia.

"The changes are important because the Bering is rich with pollock, salmon, halibut and crab, already yielding nearly half of America's seafood catch and a third of Russia's," wrote the *New York Times*.

"Recent studies have projected that in a few decades, there could be lucrative fishing grounds in waters that were largely untouched throughout human history."

The resulting dispute between Russia and the U.S. today is as much about fishing and huge oil potential in the Bering and Chukchi Seas as it is about outgrown historic treaties and Cold War politics.

The Bering Sea is the most northerly part of the Pacific Ocean, dividing Alaska and Siberia. When the U.S. bought Alaska from the Russian Empire in 1867, it took over control of the Bering Sea from the Russians.

At the time, the Alaska Purchase was criticized by the American public and considered a financial burden, given Alaska's remote location and the difficulty of governing its indigenous population. But it turned out to be a sweet deal for the Americans, giving them access to a rich bounty of fish, seals and mineral resources, and a geographical advantage during the Cold War.

Why, then, did the Russians sell off "Russian America" and its future wealth?

In 1867, Russia was financially strapped and feared Alaska would fall fast in any future conflict with its Crimean War rival, Great Britain. Rather than risk an embarrassing loss, the Russians sold off the land to the Americans for $7.2 million US, according to the *Encyclopedia Britannica*.

Near the Cold War's end, the two countries signed a treaty dividing up the sea between Alaska and Siberia, but the U.S.S.R. wouldn't ratify it, citing that the U.S. had taken 50,000 square kilometres away from its neighbour.

In 2002, a group of Russian senators met to discuss plans to seek a renegotiation of the 1990 treaty with the U.S. and compensation for $200 million a year in lost fishing revenues.

"In 1990, the treaty was seen as a breakthrough in Soviet-American relations. It was an unconsidered measure designed to bolster a dialogue that was just starting. Today, under different political circumstances, we can return to this problem," said Alexsandr Nazarov, chairman of the Russian council's Committee for the North.

Russia is claiming it lost out on 1.6 to 2 million tons of fish in areas of the Bering Sea where it has no maritime rights, reported the journal *Russia in Global Affairs*.

That's more than $2 billion US—and still counting. No agreement has been signed yet, temperatures are rising and the fish are still fleeing north, seeking the cooler waters near the last Arctic ice floes.

Race to the Polar Pinnacle

No one officially "owns" the North Pole. In fact, the home of Santa Claus himself does not even exist (don't tell the kids) but is an imaginary point along the earth's northern axis where lines of longitude converge like the slices of an orange.

Unlike its polar opposite, the South Pole—which is located on a landmass known as the continent of Antarctica—the North Pole cannot be definitively pinpointed because it lies amid ice covering the Arctic Ocean that is always shifting location with the wind and current.

The closest land to this elusive pinnacle is the north coast of Canada's Ellesmere Island and Denmark's Greenland, which are both about 725 kilometres away.

Real or not, the North Pole is another political hot spot in the global race to stake claim to the Arctic seabed and its buried riches as climate change opens up the area to travel and exploration.

Three Arctic nations—Canada, Russia and Denmark—want to grant the jolly, old fellow national citizenship. Under the Law of the Sea Convention, a nation has exclusive economic rights to the sea around its coastline up to 200 nautical miles (370 kilometres). A country can extend this to 350 nautical miles (648 kilometres) if it can prove the geological make of its landmass matches the geology of an extended underwater shelf.

"The shelf is the relatively shallow extension of a landmass to the point where the bottom drops into the oceanic abyss. But in many places, the drop-off is a gentle slope or is connected to long-submerged ridges that, if precisely mapped, might add thousands of square miles to a country's exploitable seabed," wrote the *New York Times*.

The U.S. Geological Survey estimates that 412 billion barrels of oil and gas lay beneath the Arctic ice, or one-quarter of the world's last untapped deposits.

Companies such as ExxonMobil, Shell, Gazprom and other energy producers have already found 400 oil and gas fields that hold the equivalent of 240 billion barrels. Most of these discoveries don't yet have the pipeline or shipping infrastructure in place to deliver to a hungry marketplace.

"Although much is made of the presumed riches under northern waters, jurisdiction over the pole is unlikely to bring any kind of resource bonanza with it," wrote the Canadian Press (CP).

It may not really matter who gets the North Pole, said Michael Byers, a global politics and international law expert with the University of British Columbia.

"Because the North Pole is so far away, and the prospect of significant hydrocarbon reserves at the North Pole are so insignificant...the only value is symbolic," Byers told *CBC News* on March 12, 2009.

It's the Lomonosov Ridge that all three countries have their underwater sonars and sensors pointed at. The 1963-kilometre-long undersea mountain range runs right beneath the North Pole, wrapping around the top of the world from Siberia to Ellesmere Island and Greenland.

Experts say the ridge has 10 billion tonnes of oil and gas and significant amounts of gold, diamond, tin, manganese, nickel, lead and platinum, the UK's *Daily Mail* reported.

The nation—or nations—to prove its country is connected to this underwater ridge will gain access to its buried treasures.

The Russians have already planted a rustproof titanium flag on the seabed below the North Pole, an act viewed as ostentatious and showy by other nations. In 2007, Russian scientists dove beneath the ice and conducted research on the Arctic seabed. They later announced they had science-based proof the Lomonosov Ridge is a "natural prolongation" of Russia's continental shelf.

"One newspaper printed a map of the 'new addition,' a triangle five times the size of Britain with twice as much oil as Saudi Arabia," wrote the *Daily Mail*. "The dramatic move provoked an international outcry. The U.S. and Canada expressed shock, and environment campaigners said it would be a disaster....Observers say the move is typical of Putin's muscle-flexing as he tries to increase Russian power."

Shortly after Russia's controversial display, Denmark joined the race to the North Pole, sending a group of 40 scientists to map the Arctic seabed. The Danes even have a Facebook page in support of Denmark's claim to the North Pole, with more than 7000 members saying, "Give the North Pole to Denmark." In a CBC interview on March 12, 2009, Ron McNab, a retired scientist who recently served on the board of the Canadian Polar Commission, stated that preliminary data from the scientists' trip shows that the North Pole falls under Danish Greenland's jurisdiction.

A week after the Danes started mapping, a Canadian turbo-charged, ski-equipped DC-3—the world's one and only—headed north with a crew and high-tech equipment to measure the subtlest of changes in the strength of the earth's gravitational field.

"Massive geological features such as mountains or undersea ridges create a slightly stronger gravitational pull. Charting those fluctuations will provide the best picture so far of the Lomonosov and Alpha ridges, which are the submerged features that define the limits of the continental shelf," wrote the Canadian Press.

The flights are meant to provide only complementary proof to support the rest of Canada's Arctic claims, which must be qualified scientifically with seabed mapping. Canada plans to send two new, recently purchased miniature subs under the Arctic ice in the spring of 2010.

Countries that joined the Law of the Sea Convention before May 13, 1999, had until May 13, 2009, to submit their claims. Other nations have a decade from the date they ratified the treaty to do so.

Not everyone agrees that mapping the ocean floor is a logical way to determine jurisdiction in the Arctic.

Ted Nield of the Geological Society in London told the *Daily Mail* that the whole process is "nonsensical."

> The notion that geological structures can somehow dictate ownership is deeply peculiar. Anyway, the Lomonosov Ridge is not part of a continental shelf—it is the point at which two ocean floor plates under the Arctic Ocean are spreading apart. It extends from Russia across to Canada, which means Canada could use the same argument and say the ridge is part of the Canadian shelf. If you take that to its logical conclusion, Canada could claim Russia and the whole of Eurasia as its own.

Seen in this light, the whole Arctic land grab scenario seems a bit foolish. After all, 75 million years ago every continent on Earth was joined together in one giant landmass. Doesn't that make us one land and one people—global citizens staking claim to world riches that already belong to us all equally?

Time, hard science and diplomatic bargaining will ultimately decide if we carve up the Arctic with median or sector lines and share the wealth—and if jolly old St. Nick gets to be a Russian, a Dane or sport a maple leaf, after all.

A Nation's Passage?

For many Canadians who have grown up or lived during one or more of the nation's heated sovereignty debates, a fierce national pride and protectionist spirit flare up whenever anyone challenges Canada's identity or autonomy.

In the mid-1990s, Canadian citizens, young and old, were perplexed and existentially challenged by the flurry of big questions being raised in the media about our national identity.

What does it mean to be Canadian? How do we define Canadian culture? Are we a "melting pot" or a "tossed salad"? A rich tapestry, culturally void or culturally distinct? A product of American pop culture—or something hybrid and blended, whose result, however, is uniquely us?

In the '90s, even the Canadian Broadcasting Corporation (CBC) was struggling to survive on a new information super-highway overrun by U.S. programming. It was confounding enough to be a teenager in the age of MTV, political correctness, Barney, 45-kilogram waif models, caller ID, boy bands, Pepsi Clear and throw-away pop stars—let alone grow up in a country that was facing an identity crisis.

We've been called "America's attic" and likened to a watery vegetable by the well-known Canadian actor Mike Myers, who said of his native land, "Canada is the essence of not being. Not English, not American, it is the mathematic of not being. And a subtle flavour—we're more like celery as a flavour."

Italian-American gangster Al Capone joked at our expense, "I don't even know what street Canada is on."

Even Canadian media guru Marshall McLuhan—the guy who coined the phase "the medium is the message"—once referred to Canada as "the only country in the world that knows how to live without an identity."

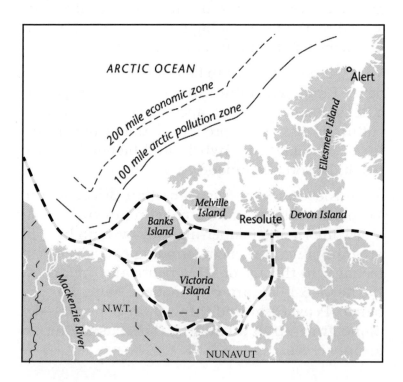

Although some have painted us as ambivalent, invisible and sponge-like, others have defended our patriotic pride.

British PM Sir Winston Churchill said, "There are no limits to the majestic future which lies before the mighty expanse of Canada" and former Montreal Alouettes star kicker Don Sweet simply exclaimed, "Vive la Canada. This country is not for sale."

When it comes to Canada's "true north strong and free," the country has been criticized for its northern neglect, but with the stakes this high—and so many nations looking north—Canada is stepping up. Prime Minister Stephen Harper has told Canadians and the world "Canada has a choice when it comes to defending our sovereignty over the Arctic. Either we use it or lose it. Make no mistake, this government intends to use it."

Since that pivotal speech on July 9, 2007, the nation has bolstered its military, research and development efforts in the Far North.

On April 2, 2009, Minister of National Defence Peter MacKay announced that the Canadian Rangers—members of the Canadian Forces reserve who typically provide military presence in more remote, isolated areas—would begin air patrols on Ellesmere Island as part of Operation Nunalivut 2009. In the Inuktitut language, *nunalivut* means "land that is ours." MacKay said:

> The Canadian Forces play an important role in achieving our goals in the north, which is why the Government of Canada is making sure they have the tools they need to carry out a full range of tasks in the Arctic, including

surveillance, sovereignty and search-and-rescue operations.

Like the other Arctic-5 nations, Canada's main goal is to defend and extend its national territory and to secure access to the vast natural resources below the Arctic seabed.

The oil reserves in Canada's Arctic Archipelago are estimated at less than one billion barrels, but there are trillions of cubic metres of gas waiting to be harvested. No one contests Canada's rights to the land north of its mainland, so its claims on this front are secure. It's Canada's famed waterways that are still in question.

As discussed in Chapter One, the melting Northwest Passage could provide a year-round shortcut to Asia by 2020. Naturally, other countries want to use this route, but Canada says it has exclusive rights to these waters because they're internal and can therefore regulate traffic and pollution program checks at will. However, the U.S., EU and other international parties do not recognize Canada's claim and regard the passage as an international strait, open to all.

In October 2009, Conservative MP Daryl Kramp introduced a proposal in the House of Commons to rename the Northwest Passage the "Canadian Northwest Passage" to symbolically bolster Canada's sovereignty over the disputed waterway. The idea was enthusiastically backed by Liberal, NDP and Bloc MPs, a rare moment of party unanimity in Canadian politics, and a swift resolution on the issue was expected.

Soon after, Inuit leader Paul Kaludjak addressed MPs in the House, asking them to reflect Canada's

Aboriginal culture in any renaming of the Northwest Passage.

Kuladjak told the House that the Northwest Passage is already known to many Inuit as "Tallurutik," an Inuktitut name that refers to an Inuit tattooing ritual and a related landscape feature on Devon Island, at the eastern mouth of the passage, which looks like tattoos on a woman's chin. "Talluq is a chin in Inuktitut and tattoos on the chin of a woman were called tallurutiit," said Kuladjak. "That's where the name comes from. The elders said that from a distance, you can easily see this."

When the House reconvened on the issue in late November, the discussion derailed as Liberal MPs withdrew their support, arguing that Inuit leaders were not fairly consulted on the issue. Nunavut's land-claim agency, Nunavut Tunngavik Inc., objected to the name "Canadian Northwest Passage," citing they have provisional rights, under the federal act that created Nunavut in 1999, to review all proposals to change the names of geographic features in the Canadian North.

By meeting's end, no name had been agreed upon. Another round of discussions was scheduled for December to consider the revised plan of a hybrid name that combines "Canadian Northwest Passage" with an Inuktitut word.

University of British Columbia Professor Michael Byers argues that renaming the Northwest Passage to fortify Canada's sovereignty claims to the waterway is not only unnecessary, but could damage the country's legal position in the long run.

The international law expert told Canwest News Service that the proposed name change "accomplishes nothing—apart from revealing insecurity about the strength of our legal claim."

Byers says that Canada's sovereignty over the passage is already secure, despite claims by the U.S., the European Union and other countries that it's an "international strait." However, if the dispute ever went to an international tribunal, Canada may face a maelstrom of scrutiny for its revealed insecurity.

"If the United States was litigating with Canada over the Northwest Passage, I'd expect them to point to any evidence that the Canadian government

wasn't completely convinced by its own position," said Byers. "The proposal to change the name is clearly aimed at strengthening Canada's position, which suggests that Parliament feels it needs strengthening. As a result, the motion could have the opposite effect from what its proponents intend."

Time will tell, how this long-standing dispute unfolds. As the issues come under review, what will it mean for Canadian sovereignty and stewardship in the Far North? Will the icy peaks of our northern identity melt under the spotlight of lawful scrutiny or rise to the task at hand?

Canada is waving its flag, but no one's watching—and perhaps for good reason, says a defence expert. J.L. Granatstein, a senior researcher at the Canadian Defence and Foreign Affairs Institute, wrote in the *Globe & Mail*:

> To concede that Canada controls [the Northwest Passage] could have implications on the other side of the globe, and seafaring states are fearful of a precedent that might let less responsible countries than Canada close off or seek to exercise control over international sea routes. Pressing for a definitive resolution of the question, in other words, might not result in a decision that pleases Canadians.

Does the Northwest Passage really matter? While it may seem countercultural or even outright unpatriotic to Canadians to even contemplate such a question, the reality is that what Canada perceives as its greatest sovereignty claim in the Arctic may, in the long term, be moot. In short, some argue, it may be more "Canadian" to set aside domestic interests and consider the bigger picture.

As the ice recedes north of the Northwest Passage, an even shorter, cheaper route will open up across the Arctic Ocean. At this point, foreign ships won't even bother with the longer, meandering journey through Canada's national waterway.

Granatstein suggests that Canada toe its current line and wait this one out:

> Canada should continue its present largely administrative approach to the passage in dealing with the United States and other maritime powers for the short term. Why wave an emotional flag about an issue that will soon largely disappear? Why even consider [entering a legal battle] if we might lose? So long as Canada can enforce its pollution controls, what else matters?

In the meantime, Canada and the U.S. are agreeing to disagree on the matter. Prime Minister Stephen Harper says his government will continue to build Canada's strength and sovereignty in the north through military, economic, social and environmental initiatives. Together, the U.S. and Canada have begun a "clean energy dialogue" and will collaborate on energy research related to advanced biofuels and energy efficiency.

Shortly before Harper's Ottawa meeting with U.S. President Barack Obama on February 19, 2009, *Embassy Magazine* wrote that despite Bush's formal go-ahead on the Arctic before leaving office, northern issues aren't currently high on the American agenda:

Aside from the region's energy potential, the Arctic is undoubtedly far down on Mr. Obama's list of priorities. Likewise, Mr. Harper will likely have more important things to discuss during his first meeting with the leader of the free world than this contentious bilateral issue. Still, as climate change continues to free up the Arctic, shipping will increase and concerns over security and further environmental damage will become more pressing. Canada and the U.S. will have to sort out where they stand and how they can work together.

Canada stands firm in its claim that the Northwest Passage is its own. The U.S. and other nations disagree. For now, everyone is treading water on the issue. Some experts say Canada's "pivotal" Arctic sovereignty battle may not be worth fighting after all—that, in this case, it would better serve Canada, the U.S. and the free world to think globally rather than unilaterally and nationally.

So, who's north is it, anyway? We will see, as new fences are drawn around Hans Island, the North Pole, the Northwest Passage and in the Beaufort, Barents and Bering seas—at the same time, mapping new diplomatic relationships in tomorrow's circumpolar neighbourhood.

CHAPTER THREE

Global Warming: In Our Lifetime or the Day After Tomorrow?

In the 1994 apocalyptic blockbuster film, *The Day After Tomorrow*, the world is thrown into chaos by extreme and sudden weather changes brought on by global warming.

Baseball-size hailstones pummel Tokyo. Tornados tear up Los Angeles. Snowstorms cripple London and New Delhi. Manhattan is flooded by a tidal wave then flash-frozen within a few hours. A new ice age has begun.

This chilling doomsday scenario, parts of which were deemed "sensational" and "misleading" by scientists, blames rampant industry and prolonged political apathy for destroying our planet and hurling modern civilization into a human-induced deep freeze.

Viewers are left with the unsettling question of whether the film's "intense scenes of peril" accurately depict the possible scale of a real-life cataclysm, and if so, to what extent.

Not to mention a puzzling paradox: rising temperatures and plunging mercury.

Isn't global warming about heating things up, not cooling them down? Do blizzards in Vancouver and heat waves in Edmonton signal of more freak weather to come? Could a world out of balance really boot us back into another ice age? If so, how—and, more importantly, when?

The answers to these questions are complex, like the planetary forces at play. First of all, any scientist who is not being paid to ignore the facts on global warming would impress upon others that climate change is not just imminent, but happening now.

The Intergovernmental Panel on Climate Change says, "Scientific evidence for warming of the climate system is unequivocal."

This notion is backed by the National Oceanic and Atmospheric Administration (NOAA), which stated, "There is a better scientific consensus on this issue than any other...with the possible exception of Newton's Law of Dynamics."

So unless you're gravitationally challenged, the question to ask is not whether global warming is real, but to what extent is humankind's collective footprint is disrupting our planet's natural cycles.

Mainstream news headlines in recent years have captured the mounting severity and scale of global weather events that most experts correlate with climate change:

"Cyclone Kills Hundreds in Bangladesh."
"Rising Sea Floods Indonesian Capital."
"Death Toll Rises in Worst Tibet Snowstorm

on Record." "Warming World to Fan More Australian Wildfires." "Record Drought Cripples Life Along the Amazon." "Worst Heat Wave in 100 Years." "Hurricane Katrina Pummels Three States." "Earth on the Brink of an Ice Age."

Environmental scientists read the news and go to work analyzing these catastrophic events in the hopes of better understanding them. Political activists point fingers in an effort to spur debate and convey a sense of urgency around the issue.

Meanwhile, skeptics tend to believe global warming is a force of nature, not caused by humans. They find scientific assessments faulty or inconclusive, or simply cultivate the appropriate rhetoric and beliefs to protect and promote their own interests.

In 2004, *Science* magazine analyzed almost 1000 peer-reviewed scientific papers on global warming, published over an entire decade, in the hopes of wiping out the lingering myth that scientists disagree about the reality of global warming. Peer-review, or "refereeing," is the process of subjecting your ideas or research to the scrutiny of other experts in the same field of study to ensure or enhance the quality of scholarly work being carried out. The system is used in science, law, medicine, accounting and many technological fields.

Science magazine's findings were conclusive: not a single paper challenged the consensus among scientists that humans are causing the earth's temperature to rise, and that it's a significant problem of our times.

Another later study conducted in the media had a much different result. The findings revealed that 53 percent of mainstream news articles gave equal weight to the claims made in so-called "scientific articles." Translation: not only does the media frequently fail to discriminate enough with its sources—in this case, by failing to use peer-reviewed papers for its own study—but it often gives disproportionate airtime to the most sensational new stories of the day. In the name of ratings, the media delivers "infotainment," often in place of solid fact. Its study, unfortunately, was likely more widely read than *Science* magazine's.

"The misconception that there is disagreement about the science has been deliberately created by a relatively small number of people," says American environmental activist and former U.S. vice-president Al Gore, in the Academy-Award winning documentary film, *An Inconvenient Truth*, directed by Davis Guggenheim.

The $49 million box-office hit exposes the denial perpetuated by government and big business around global warming, while it strives to educate lay people on the known facts of climate change.

In the film, Gore relates the story of a media-leaked memo originating from a special interest oil and gas group. The memo stated that the group's objective was to "reposition global warming as a theory rather than a fact."

He compares this memo to a similar memo from Brown & Williamson Tobacco Company (B&W) in the 1960s, leaked just after the surgeon general drew a definitive link between cigarette smoking and lung cancer.

The B&W memo reportedly stated, "Doubt is our product, since it is the best means of competing with the 'body of fact' that exists in the mind of the general public. It is also the means of establishing controversy."

Clearly, the hottest ecological debate of the 21st century is anything but simple, with competing interests from science, politics and business adding to the discussion.

The lay Canadian is faced with the unenviable challenge of sifting through weighty science, blockbuster rhetoric and spin-doctoring to determine the truth about a controversial subject that's rapidly rewriting our science textbooks.

So what do we actually know about climate change? And what does this complex and on-going debate have to do with Arctic sovereignty and Canada's roles and responsibilities in the north?

Research shows that in just over a century, modern industry has tipped the scales of our planet's ecological balance. Dramatic changes are unfolding on the earth that haven't been seen for thousands of years—since the end of the last ice age.

We're seeing shifts in weather patterns and temperatures—melting glaciers, rising sea levels, floods, droughts and heat waves—changes in plant and animal diversity, species migration and disease outbreaks, none of which can be explained by charted natural cycles through time.

Environmental anomalies are showing up everywhere, from the High Arctic and Asia to the lowlands of Africa and Antarctica.

The facts point to a relatively short time period of dramatic change—from the late 18th century onward—when the wheels of modernity cranked into motion. The Industrial Revolution saw a sharp rise in the burning of fossil fuels and deforestation. Large quantities of carbon dioxide (CO_2) were dumped into the atmosphere, spurring changes in the modern climate, a phenomenon widely known as climate change.

The wheels haven't stopped turning since. In fact, they've accelerated. As Winston Churchill warned in the '30s when Europe was hit by record-breaking storm conditions, "We are entering a period of consequences."

A look at the basic science of global warming provides a starting point for understanding the significance of the changes taking place, what they mean for the Arctic and where our planet could be headed.

For example, should we a) stockpile sunscreen, bottled water and peanut butter; b) invest in scuba gear for an exceptionally rainy day; c) brace for life—or death—at sub-zero in a throwback to the Pleistocene Era; d) all of the above; or e) just do nothing? After all, global warming is a "scam perpetrated by scientists with vested interests, but in need of crash courses in geology, logic and the philosophy of science."

Those words were written by British geologist Dr. Martin Keeley in an article published by the

BBC, which, incidentally, did not contain a single reference to his personal stakes in the oil and gas industry.

It's a well-known fact that we rely on the sun's radiation to stay warm. So we'll use simple science as the building blocks to construct a view of the bigger picture.

During the day, the earth absorbs solar heat. Some gets radiated back into space, but the earth's atmosphere—which is made up of heat-absorbing gases—acts like a blanket, trapping some heat in. This natural process, called the "greenhouse effect," is what keeps us warm at night.

If not for natural greenhouse gases like carbon dioxide and methane, the earth's temperature would drop by more than 30° C—to about -18° C—when the sun sets.

As it is, the earth's surface is kept at a comfortable global average of 15° C, making it fit for the survival of humans, animals, plants and other organisms.

Here's the problem: carbon dioxide is also the main greenhouse gas driving global warming. Excess amounts of it are produced through deforestation and the combustion of fossil fuels such as coal, petroleum and natural gas. As CO_2 accumulates in the earth's atmosphere, more of the sun's radiant heat is trapped. This intensifies the greenhouse effect—good for prize-winning tomatoes and bikini-clad sun worshippers; not so good for the equilibrium of the planet.

Spurred on by the desire to know if a drop in CO_2 caused the ice ages, late 19th-century Swedish chemist Svante Arrhenius studied the relationship between CO_2 levels and the earth's temperature. What he found took him off his original course but put his name in the science history books.

Scientists have since learned what Arrhenius wanted to know—that a dramatic shift from hot to cold can, in fact, take place, and within just a decade. By studying northern ice core samples—which contain ice, snow and trapped air from another time period—researchers determined that toward the end of the last ice age—about 13,000 years ago—the climate on the earth suddenly did an about-face.

The chain of events went something like this: northern ice sheets melted and flooded the Atlantic ocean, which changed salt concentrations and disrupted the Gulf Stream ocean currents that warm America and Europe. This rapid chain of ecological events plunged the northern continents back into an ice age, called the Younger Dryas, for 1300 years.

So, although *The Day After Tomorrow* may seem like hokey science fiction, experts now know it's possible that a rise in carbon dioxide levels over a few decades could suddenly—nature willing—flip a switch and shut off the heat pump that regulates our northern climate.

As for Arrhenius, his research has helped us understand the complex relationship between temperature and greenhouse gases. Advancing the work done by Irish physicist John Tyndall, who theorized that a connection existed between rising greenhouse gas levels and climate change, Arrhenius quantified Tyndall's hypotheses. His pen-and-paper calculations revealed what would happen to average global temperatures if CO_2 levels were doubled or halved.

His findings were as follows: if CO_2 levels doubled, the earth's temperatures on average would rise by 5° to 6° C. Conversely, if CO_2 concentrations thinned by half, average temperatures would drop 5° to 6° C. Today's state-of-the-art climate models give a range of between 2° and 4.5° C, so Arrhenius and his long-hand mathematics weren't far off.

Arrhenius, however, is best known for the conceptual link he drew between industrialization and climate change. He looked at the dark clouds of progress pouring from factories, power plants and locomotives and saw that fossil fuel consumption would gradually warm up the earth. But Arrhenius theorized it would take thousands of years—an assumption that proved to be quite wrong.

It was a chemist named Charles David Keeling who decided, in the 1950s, to measure CO_2 levels on the earth. He worked from the middle of the Pacific Ocean at the Mauna Kea Observatory on Mauna Kea, Hawaii. His painstakingly charted observations were compelling.

For two years, he tracked the seasonal rise and fall of CO_2 levels. It became clear that in spring and summer, trees in the Northern Hemisphere inhaled CO_2 for photosynthesis, lowering CO_2 levels in the atmosphere (most of the world's forests are north of the equator). Conversely, in autumn and winter, the trees exhaled and dropped their leaves, sending CO_2 levels up again.

By 1961, three years after he started taking his meticulous notes, another, more alarming—and unnatural—trend emerged: a steady overall rising of

CO_2 levels. When applied to Arrhenius's observations of the relationship between CO_2 and heat, it was clear: climate change had begun.

Carbon dioxide concentrations in the atmosphere are measured on Keeling's now-famous graph—the "Keeling Curve"—in parts per million (ppm). In 1959, the CO_2 level was recorded at 316 ppm. Two years later, it had risen to 317 ppm.

To put that into perspective, in more than 650,000 years—and several ice ages—the concentration of CO_2 in the earth's atmosphere never approached 300 parts per million. Prior to the Industrial Revolution, carbon dioxide in the earth's atmosphere was about 280 ppm.

By 1975, CO_2 levels had jumped to 330 ppm, and by the end of the booming 1980s, they'd hit 353 ppm.

The NOAA continues to track carbon dioxide levels from Mauna Kea. The trend that was pinpointed almost a half-decade ago has persisted. Today's CO_2 levels are roughly 386 ppm, and they are expected to rise to 500 ppm—double pre-Industrial levels—by the middle of the century.

The Arctic has always been frozen, except for a prehistoric period 55 million years ago when massive amounts of greenhouse gases were released into the earth's atmosphere. During this time, the Arctic was warm, wet and ice-free, experts say.

Modern climate change is once again melting the Arctic. In this new northern frontier, issues of science and politics are virtually indivisible. The Arctic's strategic location and massive resource wealth are making the north a strong focus of national interest. New, coveted fishing lanes and billions of barrels of now-reachable oil and gas, combined with an ill-defined picture of state ownership rights, are a recipe for uncertainty and possible danger.

As ice turns to water, and nations scramble for territory and profits in the new Arctic, the rules are unclear. Current legal and political structures aren't designed to referee unresolved sovereignty disputes or break up diplomatic gridlock.

Meanwhile, Mother Nature is hard-pressed to keep pace with the rampant modern development that threatens to cause wide-scale environmental fallout. Although plant and tree growth should increase with higher CO_2 levels—helping to reduce overall atmospheric CO_2 and increase oxygen—it is unlikely to control or reverse climate change because vegetation on the earth is decreasing with human expansion, as people clear-cut for crops and living space.

The science is abundantly clear: our beautiful blue and green planet is heating up faster than at any other time in recorded history.

What was believed, in the '80s, to be a "slow-motion catastrophe" that would not take effect for generations has gained momentum at an alarming rate, says famed environmentalist David Suzuki, a long-time geneticist, activist and host of *The Nature of Things*.

Climate change has already upset the natural balance of the earth's ecology. What remains to be seen is how she will respond and what government and industry will do to protect her.

Says Suzuki, "Yet here we are playing Russian roulette with features of the planet's atmosphere that will profoundly impact generations to come. How long are we willing to gamble?"

Polar Bears on Thin Ice

A fierce storm whipped across the northern coast of Alaska in autumn of 2004. In its wake, field researchers found the carcasses of four dead polar bears in one month, floating in the same part of the sea.

The discovery led scientists to look more closely at how High Arctic climate change is threatening the livelihood of a species that depends on ice as a platform to hunt seals, rest and rear its young.

Arctic sea ice has shrunk dramatically in recent years, according to climate data gathered by the National Snow and Ice Data Center.

With average autumn temperatures rising in the Arctic—5° C above normal by 2008—sea ice coverage has hit a new record minimum of 1.6 million square kilometres lower than measurements recorded between 1979 and 2000.

The main ice pack is now a full 480 kilometres off the northern coast of Alaska. To the hungry polar bear, that means staying on land—where it can't hunt seals—or travelling greater distances across water to the nearest ice, to catch seals through air holes carved into the ice.

This decision is often one between life or death, sink or swim. The polar bear, while adapted to tread water close to the shoreline, can easily cross 24 kilometres of water to find food or shelter. But scientists think the four that succumbed to battering waves, exhaustion and hypothermic conditions likely travelled 96 kilometres or more in search of a suitable ice floe.

In all likelihood, they weren't alone. Before the deadly storm hit, researchers spotted 51 polar bears swimming in the same vicinity, a number of which may also have drowned in the squally seas. As temperatures climb and ice recedes, the number of polar bear casualties is expected to rise.

In the not-so-distant future, these grand, white bears of the Far North could disappear entirely, along with the waning ice-blue pinnacles that dart the changing Arctic landscape.

Fewer than 25,000 polar bears remain in the wild, says the World Wildlife Fund, and some populations face extinction within the century, including Canada's Hudson Bay polar bears.

This population reportedly fell 22 percent from 1194 in 1987 to 935 in 2004, according to a study by the U.S. Geological Survey and the Canadian Wildlife Service. It's estimated that two-thirds of all polar bears could die off by 2050.

Ice is melting much more rapidly than the majority of climate models projected it would by mid-century. If this trend continues, many scientists predict we'll have an ice-free Arctic during summer as early as 2012.

Environmental lobbying led to the polar bear being listed as a threatened species under the Endangered Species Act in May of 2008. The state of Alaska filed a lawsuit in U.S. district court to reverse the decision. Local government claimed that sufficient measures were already in place under the Marine Mammal Protection Act to manage polar bear conservation.

A terse warning came in a state-issued press release by Department of Natural Resources Commissioner Tom Irwin:

> Inappropriate implementation of this listing decision could result in widespread social and economic impacts, including increased power costs and further increases in fuel prices, without providing any more protection for the species.

Later that year, a settlement was reached that required the federal government to designate "critical habitat" for polar bears in ocean waters off the Alaskan coast.

This move is expected to restrict any oil and gas activity that could pose a threat to polar bear populations.

"Other than global warming, the worst thing that's going on in polar bear habitat right now is oil development and the potential for oil spills," said Kassie Siegel with the Center for Biological Diversity, one of the three groups—including Greenpeace and the Natural Resources Defense Council—that forced the ruling, according to *Anchorage Daily News* reports.

With polar bears already on thin ice and Arctic thaw gaining slippery momentum, just where this "critical habitat" will be located remains to be seen.

Ultimately, the ice will vanish, and remaining polar bear and seal populations will face survival in an environment they are ill-adapted to live in.

As both industry and global warming advance, what, in the end, will drive the polar bear to its final frontier—extinction?

It's a slippery debate, one in which the ecological footprint of industry strides in step with the bottom line, not nature.

Darwin's Moth Scores One for Nature

Hollywood goes weak in the knees for a story of good against evil. What could be more menacing than power-hungry humans who are destroying the life-giving planet Earth?

Since the mid '90s, a string of global-warming flicks—*The Day After Tomorrow, Water World, A.I. Artificial Intelligence, An Inconvenient Truth, The Arrival*—has hit the silver screen.

Even the megahit *Lord of the Rings*, in true Tolkien style, weaves an environmental theme through its fantasy epic—equating industrialization and pollution to the forces of evil, and living in harmony with nature to the forces of good.

Science and fantasy overlap in *Fellowship of the Ring*, in a doubly symbolic scene that depicts a tower-imprisoned Gandalf the Grey rescued by a delicate white moth.

The moth—an ancient Greek symbol of peace and hope—has gained new significance to evolutionary scientists since the mid-1950s, when the winged insect became proof of the genetic consequences of human-made change.

The smoky, sooty tale of "Darwin's moth" is a testament to nature's adaptive genius (so much so that bible-thumpers tried to get it wiped from biology textbooks). In 19th-century England, the peppered moth (*Biston betularia*), was a predominantly white-winged moth lightly speckled with black; some had dark, almost black bodies, but they were extremely rare.

This beguiling moth sought camouflage from birds in the lichen growing on trees. As expanding industry belched out smoke, parts of Britain became quite polluted. The tree lichen died and, in its place, the bark was coated by black residue. The white moth, suddenly very conspicuous, fell prey to birds, while its rarer black relatives survived and reproduced. By 1895, 95 percent of peppered moths were black. The insects adapted genetically to industrial pollution, and natural selection changed the colour of Britain's moth population.

Then, just as rapidly, natural selection changed things back. Britain's Clean Air Act was passed in 1956 to address the smog problem and, sure enough, lichen began growing on trees again. The black moth population dropped off and the white variety flourished as before.

Another chain reaction—on a wider scale—is occurring in the Arctic. The northern climate is warming twice as fast as the rest of the world. Consequently, sea ice has declined dramatically over the past 30 years, according to NSIDC statistics.

As the ocean warms, the cycle repeats: more ice melts, water levels rise and temperatures follow. Like a ball accelerating downhill, the Arctic is heating up and gaining momentum as it goes.

This uninterrupted cycle of heat and thaw is driving a trend toward warmer temperatures around the globe. The Arctic melt is wreaking havoc on regional and global weather systems and is dramatically affecting the earth's biodiversity—in ways not unlike Darwin's moth.

Scientists have identified species that have changed their genetic makeup to cope with global warming. In the past, animals have shown flexibility—or plasticity—in adapting their behaviour or response to new environmental forces. However, the following examples show a deeper, more permanent change on the level of genetic coding.

The mosquito *Wyeomyia smithii* now goes dormant for the winter 8 to 10 days later than it did in the 1970s. Because the insect does not respond to hot or cold, scientists know this change is not an environmental response, but is actually genetic.

The non-biting insect is found from Mexico to Canada and lives in the purple pitcher plant, where it stores its food for winter. As the climate warms, those mosquitoes that start hibernating later—because of a genetic variation—have more time than normal to gather reserves and, as such, have an increasingly better chance of surviving winter.

Mosquitoes that enjoy this selective advantage pass their late-hibernating trait on to their offspring. The process repeats itself until most mosquitoes enter dormancy at the best time to ensure their survival through winter.

"Rapid climate change is now actually driving the evolution of animals—that is a dramatic event," said Dr. Christina Holzapfel, who, with her husband, Dr. William Bradshaw, conducted the lab study at the University of Oregon.

In Canada, the red squirrel is another species that is evolving because of global warming. Warmer springs bring an abundance of spruce cones on which to feed, and the females that breed sooner give a head-start to their offspring. The young squirrels are bigger, stronger and more self-sufficient when autumn arrives and it's time to store food for winter, says Dr. Stan Boutin, the researcher from the University of Alberta's Department of Biological Sciences, who led the study.

Those animals that can't adapt quickly enough are left to cope and find their new place amid the dynamic shifts taking place, or else face extinction.

The length of seasons has been changing since the 1970s, which is disrupting natural animal habitat ranges, distribution and migration and breeding patterns.

Species are now found in places they've never lived before. As the Arctic warms and species move farther north, new biological communities are created that lead to competition for food and new interactions, which drive further changes in the natural ecosystem.

Cold-dwellers are running out of places to retreat. Those who've extended to the far northern reaches face the prospect of dying off as environmental conditions change and can no longer support them.

The evidence is conclusive: the world's climate and biosphere are rapidly changing. These changes are unparalleled in recent millennia.

While historical climate change has seen species evolving or being wiped out many times over—albeit over hundreds of thousands of years—the turbo-speed of human-powered global warming is raising the spectre of runaway consequences in the Arctic and around the world.

Scientists say survival tends to favour the smaller animals—insects, birds, squirrels—whose rapid life cycles allow them to adapt genetically quicker than larger animals and species that live and breed over longer periods.

As rodents, insects and birds carrying diseases thrive under selective advantage, scientists believe they could become more dangerous to humans because global warming affects the range and nature of infectious diseases.

The Arctic is changing its northern face: temperatures are rising, sea ice is melting, shipping is increasing and oil and gas activity are heating up. All told, these events are likely to have wide-reaching impacts on regional wildlife—including polar bears and ice-dependent seals—and the health and well-being of indigenous peoples who live closely with nature and rely on these animals as food sources.

Expanding oil and gas activities in the Arctic could bring a new wave of environmental stresses and pollution to compound global warming.

Some experts believe that as temperatures climb, new tree growth in the Arctic could help reduce and stabilize carbon dioxide levels and provide new industry and employment opportunities for northerners.

However, this growth could also throw off the dynamics of existing habitat, and such events as forest fires and insect outbreaks could further offset these benefits, according to the 2004 Arctic Climate Impact Assessment report (ACIA).

Around the world, ocean levels are rising twice as fast as they were predicted to 150 years ago—almost 2 millimetres per year compared to 1 millimetre annually for the past several thousand years—and weather patterns have become more severe, unpredictable and catastrophic. There is international scientific consensus that the increased burning of fossil fuels and rising carbon dioxide levels are the major culprits.

NASA is tracking environmental changes from space and reports that global temperatures may be within 1° C of the maximum temperatures of the past million years.

"That means that further global warming of 1° C defines a critical level. If warming is kept less than that, effects of global warming may be relatively manageable," says Names Hansel of NASA's Goddard Institute for Space Studies.

> But if further global warming reaches 2° or 3° C, we will likely see changes that make Earth a different planet than the one we know. The last time it was that warm was in the middle Pliocene, about three million years ago, when sea level was estimated to have been about 25 metres (80 feet) higher than today.

Such predictions make apocalyptic images of Manhattan underwater strike awfully close to home. What would we do? How would we evolve to cope with such wide-scale calamity?

Papua New Guinea's Carteret islanders may be the world's first refugees displaced by climate change. Their island in the South Pacific is disappearing under water, and people are already being moved to a larger island nearby. It's expected that up to 11,600 people may have to abandon their homes, says president Anote Tong of Kiribati, a nation made up of 33 low-lying islands in the South Pacific and home to about 100,000 people.

Already, two uninhabited islands have been swallowed by the rising tide, and another island no longer grows coconuts because of the effect of high concentrations of ocean saltwater, according to the South Pacific Regional Environment Programme.

In the Netherlands, flooding is an ongoing problem expected to intensify with global warming. A full quarter of the country sits below sea level, so the Dutch people are used to water-related issues and are savvy problem solvers. Another quarter of the country is low-lying and susceptible to flooding, so the country has worked to "reclaim" it through an elaborate water management system that includes kilometres of dunes, dikes, holding ponds and pumping systems.

High-risk areas of the Netherlands have also been selectively abandoned or "given back to the ocean," and what's known as a "flood market" has emerged—the first amphibious community of floating houses, built in the town of Maasbommel.

Today, many Dutch people believe that, in the future, citizens of the world will live in floating homes that rise and fall with the tide and the prevailing whim of Mother Nature.

What can we—our governments, communities, science and industry—do to address global warming and slow or reverse its momentum in the Arctic and around the globe?

Time is critical; while we mobilize efforts to deal with this complex issue, climate change keeps happening, and we're forced to adapt to its consequences in the process. All parties must participate.

In the mid-'90s, most countries joined the United Nations Framework Convention on Climate Change (UNFCC). The Kyoto Protocol came into effect in 2005—the first, and only, legally binding international agreement to combat climate change. To date, 37 industrialized countries and the EU have ratified the agreement to reduce greenhouse gas emissions between 2008 and 2012 by an average of 5 percent against 1990 levels.

The United States—which contributes roughly one-quarter of the world's carbon emissions—has opted out of this milestone agreement.

Because the majority of the world's carbon dioxide emissions so far have been produced by industrialized nations, Kyoto asks them to carry a greater burden than developing nations that have signed the agreement.

However, as highly populated nations—such as China and India—join the ranks of industrial progress, the race is on to find new, "green," cost-effective ways to do old, polluting activities; the alternative is to face a magnified human-made global eco-crisis.

Although they may marvel at the vision and cooperative feat of getting so many nations on board for Kyoto, a read of the protocol's measures for addressing climate change is part ironic, part sobering to environmental purists.

Market capitalism rules—even in our collective quest to save the world from its over-consumption. Under the protocol, nations that produce high emissions can legally come in over their targets by using "carbon credits."

This hot, new, 21st-century commodity can be bought from countries that have emissions to spare, while other carbon credits can be earned by sponsoring emission reduction and sustainable development projects for private industry or industrializing nations.

By 2006, this stock exchange for pollution was already worth $30 billion USD, according to the UNFCC.

These days, with climate change on everyone's radar, a powerful governing trio has emerged in news headlines: global economics, politics and the environment.

When U.S. President Barack Obama visited Canada in February of 2009, he and Prime Minister Stephen Harper discussed the worsening global economy, shared challenges in Afghanistan, oil and gas, and the environment.

That day, a "clean energy dialogue" began between the two leaders, who made a commitment to pool research expertise and invest $2.5 billion in the development of clean-energy science and technology. Their main focus is on carbon capture and storage (CCS) technology, which uses a process called carbon sequestration to collect and bury carbon dioxide underground.

An article posted by the online newswire Science Daily dispels the idea that carbon capture is a "panacea" for global warming: "As pollution bad guys go, carbon dioxide may be the media darling, but trying to capture it and lock it away could allow other repeat offenders to go free."

CCS could help plants reduce CO_2 emissions by 70 or 80 percent, but doing so will release extra pollutants in the process, such as nitrogen oxides and sulphur oxides, which cause acid rain and water pollution and destroy the ozone layer, studies reveal.

"If mining, transportation and other supporting technologies become greener in the future, the pollution penalty for carbon sequestration would be reduced," the article stated.

Other research on the cost-benefit of CCS shows that the current technology would significantly increase environmental pollution while reducing overall air quality, according to an IPCC joint task force report, which was reviewed by both governments before dialogue began.

The report concluded that, at present, CCS has only a narrow application because the technology can only capture a small percentage of the total carbon dioxide produced in Alberta's oil sands—the source of Canada's fastest-growing greenhouse gas emissions.

Greenpeace has called the Canada-U.S. climate initiative "dishonest": "It's clear that Canada's investment in CCS is nothing more than a smokescreen to disguise the government's inaction on climate change," says Mike Hudema, an oil sands campaigner for Greenpeace, which supports the exploration of cleaner and renewable energy technologies such as wind and solar. "Instead of continuing down a path the government's own advisors are saying won't work, we should be creating the clean economy of the 21st century."

In the United States, the president has approved a cap-and-trade carbon market that requires companies to buy emission permits with a specified cap on their emission levels. To increase their allowable limit, they can turn to the carbon market to buy credits from other companies that produce less carbon.

In theory, the system is set up to reward companies that have lower greenhouse gas emissions than their permissible quota. But the system's critics say, in practice, it's just a way for high emissions producers to circumvent new clean industry standards.

Forecasts by the 2004 Arctic Climate Impact Assessment are grim and unequivocal:

> Climate changes are projected to include shifts in atmospheric and oceanic circulation patterns, an accelerating rate of sea-level rise, and wider variations in precipitation. Together these changes are projected to lead to wide-ranging consequences including significant impacts on coastal communities, animal and plant species, water resources and human health and well-being.

Most of the energy consumed to drive human progress comes from fossil fuels. As a result, rapidly rising worldwide greenhouse gas emissions show no sign of abating. As the Northwest Passage opens and several nations race for a share of the Arctic's vast resources, how will they tread in an environment already stressed out by climate change? Will an environmental conscience accompany capitalist drive to the Far North, or will the quest for the earth's last non-renewable resources supersede responsibility along the way?

Even if we started reducing greenhouse gas emissions right now, it would still take decades for CO_2 concentrations to peak and start declining, the ACIA report forewarns. "Altering the warming trend will thus be a long-term process, and the world will face some degree of climate change and its impacts for centuries."

This dark, stormy prediction comes not from Hollywood, but from a forum of the world's top scientists and participating governments from many nations, all of whom agree on the fundamentals of the problem.

The real question is not, "what will we do?" but "what are we willing to do?"

Climate change is making the polar region more desirable and accessible. As prospectors, scientists, tourists and explorers head north, it will be interesting to see how our leaders respond and guide us. The future of the Arctic, after all, could determine the future of the planet.

CHAPTER FOUR

The Race for Tomorrow's Energy: Untapped Oil North of 60

B lack gold. Bubbling and crude. Ninety billion barrels of oil lie beneath waning Arctic ice, what amounts to nearly one-quarter of the world's undiscovered energy resources.

They're the only remaining virgin stocks of non-renewable oil and gas on the earth, some of the hardest to get at—through deep ice and sea—and, by no coincidence, they won't be extracted easily or quietly, without legal squabble or political siege.

Many nations are jockeying for offshore-drilling rights and prospector's privileges to seek out and siphon the promised riches of the Arctic Klondike.

The last great "gold rush" of our industrial age is, to an over-populated and fuel-hungry world—whose appetite shows few signs of abating—the final petrol promise of our times, a profligate era that has burned fast and bright, but ultimately whose fuel lines are running dry.

Today's race for tomorrow's energy is a drop in the bucket, a hollow echo in a near-vacant well, a final "fix." Arctic reserves will only provide enough oil to supply the world for three years, or America for 12, at current consumption rates, say energy experts. They will quench global thirst for fossil fuel now, delaying the inevitable drought to follow.

The consumer motto of our times—guzzle more, faster—has been so thoroughly embraced by the industrialized masses that world leaders have an unenviable job before them: to keep supplies flowing to fuel-dependent consumers while masterminding a way out of an imminent global energy crisis.

The fact that we haven't run out yet offers a false sense of immunity to what is actually a colossal and inescapable global problem. The artful dodger in us sports rose-coloured glasses while tearing up the tracks in gas-guzzling muscle cars, pick-ups and SUVs, thinking we've got generations before the pipes run dry. After all, why face the spectre of future shortages today, when there's still gas at the pumps and unrefined oil in the ground to fuel our every need?

According to the U.S. Geological Survey, the Arctic holds beneath its ice over one-fifth—or 22 percent—of the world's undiscovered yet retrievable oil and gas reserves: 90 billion barrels of oil and over 47 trillion cubic metres of gas, or about one-third of the world's known gas reserves.

Much of the oil and gas lies within the jurisdiction of the Arctic-5 nations; the Alaskan coast holds most of the oil, while the lion's share of natural gas lies near Russia.

"There appear to be only small reserves under the unclaimed heart of the Arctic," reported Britain's *The Independent*, adding that the U.S. study points to "a very different future for one of the world's last remaining pristine and utterly unspoilt regions. If the oil is there, countries [that] own it will be very likely to seek to extract it, whatever the environmental cost."

Despite overwhelming scientific evidence that humans are spurring global warming through the burning of fossil fuels, world energy consumption is expected to rise by 50 percent between 2005 and 2030, experts predict.

Most of this growth will occur in developing nations such as China, Indonesia, North Africa and the Middle East, where oil consumption is soaring because of rapid growth in regional economies, according to the U.S. government's Energy Information Administration (EIA).

In developing nations, total energy demand is expected to rise by 85 percent by 2030, compared to 19 percent in industrialized countries such as the U.S., Canada, Japan and European Union nations. Fossil fuels—petroleum and other liquid fuels, natural gas and coal—will continue to supply most of the world's energy needs, the EIA report forecasted.

Consider for a moment that there are 86,400 seconds in a day: 24 hours times 60 minutes times 60 seconds. The world consumes 1000 barrels of oil a second, 60,000 barrels a minute, 3.6 million barrels an hour or 87 million barrels a day. That's quite the big, black, bubbling, carbon-generating fuel dependency we've got on our hands!

Just under one-quarter of this consumption comes from the U.S. alone—about 22 million barrels a day—while Americans make up only 5 percent of the world's population.

China, a rapidly developing nation and currently the second-largest energy consumer behind the U.S., increased its oil consumption to 7.1 million barrels daily in 2006, up 6.2 percent from the previous year.

The People's Republic of China is the world's most populous country with 1.3 billion people. Its energy needs are exploding as the industrializing nation moves away from bicycles and mass transit and toward the private American-style gas guzzler. With oil prices among the lowest in the world helping to drive its growth, the country is on the up-swing of its energy consumption, just as developed countries such as EU nations and Japan are starting to curb their gasoline use, deterred by steeper fuel taxes, diminishing domestic supplies and a growing sense of responsibility for the environment.

Where are Canadians in this picture? The North American nation is ranked eighth in global oil consumption and burns just over 2.3 million barrels daily, behind the U.S., China, Japan, Russia, India, Germany and Brazil. Meanwhile, the resource-rich country is the fifth-largest energy exporter worldwide, with most of its supplies sent south to serve America's massive fuel needs.

According to *Oil and Gas Journal*, Canada had 179 billion barrels of oil reserves in January of 2008, second only to Saudi Arabia. Most of these reserves are located in Alberta's oil sands and are much harder to extract and get to market than conventional crude oil. Alberta's oil sands are believed to hold an estimated 2.5 trillion barrels of oil, and other nations are literally tripping over each other to bid on the mighty cache, but most of it is unrecoverable with today's technology.

There is enormous power in having oil to sell—especially when world stocks are depleting—but oil-exporting nations also depend greatly on the buying power of their fuel-hungry customers, namely Americans. Almost every energy-producing nation in the world ships its energy products to the U.S., including Saudi Arabia, which is, to this day, the world's biggest producer and exporter of petroleum liquids and second only behind Russia in crude oil production.

The Middle Eastern state relies heavily—almost exclusively—on the oil and petrol industries. In fact, in 2006, the Saudis' oil export revenues made up about 90 percent of the country's total export earnings and state revenues and more than 40 percent of its gross domestic produce (GDP), according to the International Monetary Fund (IMF). As the world shifts from dwindling fossil fuels to new energy sources, such as wind and biofuels, Saudi Arabia will be challenged to revamp its economic strategy for the future or face large-scale market instability.

Russia, whose old Soviet Union legacy still casts a shadow on world public opinion, comes out on top, however, for current economic growth, despite today's floundering global economy. In 2007, the former communist empire boosted its GDP by 8.1 percent, blazing past the other G8 countries' average growth rates. The nation's massive growth is mainly a result of increases in oil production and higher world oil prices. Moscow is also in a pretty position when it comes to the Far North, where Russia stands to gain the most from Arctic natural gas reserves, more than 70 percent of which lie in the West Siberian, East Barents and Arctic Alaska basins.

As Russia's oil and gas potential expands, so will its power and influence—as the world witnessed in early 2009 when a pricing dispute with Kiev, a former Soviet neighbour, led Moscow to slash its gas supplies to Europe in the middle of a January cold spell. Europe is heavily dependent on Russian energy; it gets about half of its natural gas imports and almost one-third of its oil imports from Russia. About 20 percent of this gas comes from Russia via pipelines that cross Ukraine. When Moscow cut back delivery to Ukraine on New Year's Day after failing to resolve debt and pricing issues with Kiev, the rest of Europe endured sub-zero temperatures without adequate heating.

Canada's economy isn't immune to such political power play either. Its dependency on other nations for fuel—principally the U.S.—also leaves it vulnerable in the event of supply shortages. Although Canada is a major energy producer in its own right, the country exports much of its domestic stocks to the U.S. and relies heavily on foreign oil and gas imports for at-home usage.

One need only recall the bizarre sight of yellow tape strung across pumps and signs reading "out of gas" at filling stations across Alberta and BC in August of 2008. A Petro-Canada refinery malfunction in Edmonton cut crude oil production by

135,000 barrels a day, upsetting the local supply chain.

The impact wasn't just felt by consumers running on empty but also by independent gas station owners running on ever-tightening margins. As oil prices rise, pump owners—who buy high volumes of product on credit—are increasingly gouged by credit card companies, making it harder to turn a profit.

During this prophetic August drought, several fuel pumps ran out of certain grades of gasoline, and their owners began selling higher grades at lower-grade prices to avoid losing customers. Other filling stations temporarily closed.

Petro-Canada's spokesman Jon Hamilton said that the refinery's shutdown was a result of a catalytic cracking unit glitch. He compared the situation to running a doughnut factory without a deep fryer. "We just can't produce product right now," he said.

Many westerners wondered how such a fuel shortage could occur in an oil-rich province like Alberta.

Petro-Canada launched emergency distribution efforts to bring in fuel from Ontario and B.C. storage facilities, but still, between 80 and 90 gas stations were out of fuel for several days.

So why didn't Canada look to its southern neighbour for help? Canada's tough cap on sulphur and benzene levels in gasoline prevented it from importing more U.S. product to ease the shortage. Benzene, a known carcinogen, is a natural component in gasoline, diesel and aviation fuel, but it can be extracted. Americans currently rank

67 out of 100 countries on their benzene content standards, according to the International Fuel Quality Center.

To avoid this problem in the future, the industry group Canadian Independent Petroleum Marketers Association is currently lobbying for "harmonized" gas standards between Canada and the U.S. so that in the event of another emergency, Canada can look south for assistance.

The Environmental Protection Agency has called for a global reduction in the total benzene content in any gasoline produced after 2011.

So, why then could Canada not fall back on homemade fuel stocks during the days of scarcity? In short, its trade obligations under the North American Free Trade Agreement (NAFTA) prevented it for doing so.

Under NAFTA, Canada has firm quotas on the amount of oil it must ship to the U.S, so it could not dip into U.S. export reserves to ease its own domestic shortage. Over 99 percent of Canada's exports and more than two-thirds of all Canadian oil reserves are shipped to the U.S., and the number has steadily risen—70 percent in 2004, up from 60 percent in 1998 and 50 percent in 1990. It's the price Canadians are willing pay for extended access to the lucrative American market.

Petrol politics between nations will no doubt become more competitive and volatile as world fuel production slows and we approach what energy analysts call "peak oil"—the inevitable crest and decline of the world's finite oil reserves.

Make no mistake. This future is not far off. We are rapidly depleting global supplies of non-renewable fuel, and experts say we'll "top out" in just two decades. The Arctic is melting—and just in time, it appears, because the great, not-so-white north is the oil industry's final panacea—but without the silver lining. These last and least digs come with a steep price tag, relatively small margins of oil and extensive drilling risks.

There's a reason prospectors haven't looked north until now. The Arctic is like the penultimate Little Leaguer who's picked second-last for the baseball team for being weaker in stature and assets than the rest, but a step up from what remains—the bottom of the barrel.

The U.S. has a Strategic Petroleum Reserve—the world's largest stockpile of government-owned emergency crude oil—created after the 1973–74 Arab oil embargo on the U.S. doubled the price of oil overnight. The reserve currently holds 718 million barrels of oil. It sounds like a lot, but if, for example, oil production or supplies were to grind to a halt, the U.S. would have only 35 days worth of fuel to fall back on before the super nation's security and economy hit "code red."

Ironically, the less oil we have, the more we consume. As energy stocks decline, demand goes up and world oil prices continue to rise. We saw that with the spike of $35 per barrel in 2003 as a result of Asian growth and the Iraq war. EIA energy reports predict $63 per barrel by 2010, which translates into higher prices at the pump.

This chain reaction we are seeing is complex, and may be catapulted at any time by a number of equally complex factors: a demand increase, decline in production, disruption in world supplies and/or artificial price setting by governments to offset embargoes, discourage consumption or promote a switch to new alternative energies as they hit the market.

Environmental sociologist William Catton draws a strong link between bulging world populations, over-consumption and crashing energy supplies in his 1982 *New York Times*' best-seller *Overshoot: The Ecological Basis of Evolutionary Change*. In the chapter "Dependence on Phantom Carrying Capacity," he wrote:

> The faster the present generation draws down the fossil energy legacy upon which persistently exuberant lifestyles now depend, the less opportunity posterity will have to live in anything like the same way or the same numbers. Yet most contemporary political proposals for solving problems of economic stagnation or inequity amount to plans for speeding up the rate of drawdown of non-renewable resources.

In other words, the faster we use up the oil and gas, the more voraciously our governments and private enterprise stake out and bid upon remaining stocks, to feed insatiable domestic use and foreign markets, or to suspend them at will. In short, petrol-power has irresistible privileges.

But while government and industry squeeze the last drops of oil from the earth, average citizens find themselves stuck between a rock and a hard place—with higher gas bills to pay on the one hand and unaffordable clean-energy alternatives, such as electric or biofuel cars, on the other.

There's a long way to go before these new technologies are cost-effective and mainstream, in both mature and industrializing nations. So what, in the meantime, will we do when the oil runs out?

The former top dog with Shell Petroleum U.S., John Hofmeister, spoke frankly to *Newsweek* in 2008 about drilling our way out of an oil shortage crisis. "The internal combustion is a great invention that has served us well for a century, but it is time for us to move on," said Hofmeister, who retired a month later.

Surprising words for an oil guy, but he was speaking, of course, in his new role as the founder of non-profit group Citizens for Affordable Energy—not for Shell.

Hofmeister wants to encourage Americans to petition their elected officials to make fundamental changes in current U.S. energy policy. We should be off-setting high pump prices and demand with an interim boost in oil and gas production. Now, that sounds more like an oil exec talking!

Hofmeister was thinking of the Senate controversy over lifting a 27-year ban on domestic offshore oil exploration, and drilling in Alaska's eco-sensitive Arctic National Wildlife Refuge (ANWR), which contains an estimated 11 billion barrels of oil and shelters millions of migratory birds as well as polar bears, caribou and other wildlife.

At the time, President George W. Bush endorsed drilling in the refuge to secure America's "energy independence" in the future, while presidential candidate and Republican John McCain treaded carefully around ANWR, instead backing the reinstatement of offshore drilling. In a refreshing departure from the status quo, Democratic opponent and future U.S. leader Barack Obama took a terse stance on both issues, citing that drilling in ANWR would destroy an ecologically sensitive area without much economic benefit and that renewed offshore drilling would never significantly lower gas prices.

To soften the blow of shifting priorities to oil industry types, the White House asserted that oil companies "are and will remain an important part of our energy future."

Politics aside, Shell's former leader calls for a hybrid solution with overlapping strategies to address the looming energy crisis. In other words, phase oil out slowly while phasing in new, cleaner, renewable energy alternatives.

Hofmeister says that along with new fossil fuel development over the next decade, the U.S. needs to boost production of alternative biofuels—namely cellulosic ethanol, which is currently in use—from 500,000 to 2.5 million barrels a day. He estimates that as more economical and eco-friendly cars gradually replace conventional gas-guzzlers, the daily demand for oil will be reduced by 2.5 million barrels.

Summarizing his ideas, *Newsweek* wrote, "Over time, the combination of significant increases in production and significant demand destruction, 'would hit the energy future on its head,' bring down prices, create jobs in the U.S. and decrease reliance on energy imported from tyrants."

By October of 2008, one month before Obama's win, much of the U.S. ban on domestic offshore drilling had been lifted. Politicians had even begun debating the merits of "directional drilling" in ANWR, which would allow the industry to access the cache of oil from outside of the refuge using advanced drilling technology that, purportedly, wouldn't disturb the area—a bit like swiping a cookie from the cookie jar without sticking your arm in—no crumb nor fingerprint. No harm nor foul.

This drilling technique is still unproven, U.S. Interior Secretary Ken Salazar told media in March of 2009, but is open for discussion. The U.S. position, however, remains firm—the Alaska refuge "is a very special place" in need of protection.

Since taking office, President Obama has championed a "clean energy" campaign to develop renewable energy sources, such as solar, wind, biomass and electric hybrid transportation, to decrease America's dependence on fossil fuels and pave the way for a cleaner, more affordable and sustainable energy future. Time will tell what happens.

These aren't new ideas, but the spectre of a future without oil finally has world leaders paying attention. There's truth to the "taboo talk" of peak oil, demand without supply, skyrocketing prices, declining production, collapse of oil-dependent economies, resource wars and damage control. As early as the 1880s—right at the beginning of the modern oil era—there was speculation about the peaking of world oil production, but technology wasn't advanced enough to prove anything.

Later, in the 1950s, when the scientific evidence was there, nobody wanted to face it—except for a certain candid Texas geologist, Dr. Marion King Hubbert, who wasn't afraid to call a spade a spade.

A reticent dialogue finally began in response to his grim and unwelcome predictions in the spring of 1956.

The place was San Antonio, Texas, where 500 petroleum engineers of the American Petroleum Institute had congregated for a three-day meeting. On the first day of the gathering, March 7, many of the conference-goers were expecting an upbeat overview of where their industry was at—i.e., "the U.S. is still the world's most powerful petroleum producer and consumer, industry potential is vast, reserves are boundless and the future is bright, so let's enjoy the ride, after all, we're generations away from running out." Yadda, yadda.

However, that is not what they heard. Instead, the first speaker, world-renowned Houston Shell Oil geologist Dr. Hubbert, shocked the crowd back to reality, captivating all and ruffling many with his fact-based claim that the U.S. is not, after all, an indomitable oil "superpower" anymore than the world's finite petroleum reserves flow in perpetuity. In mere minutes, he shook their whole world.

Hubbert said that U.S. domestic oil production would peak in 10 to 15 years and that global production would subsequently crest and begin to decline early in the 21st century, at which point global demand for oil would forever outpace global oil production. Supply-demand margins would widen over time and ultimately devastate the world economy.

For the group, Hubbert graphed oil on a 10,000-year scale—5000 years in the past to 5000 years ahead—to show that humankind's oil consumption is "a unique event in human history, a unique event in biological history. It is non-repetitive, a blip in the span of time."

The petroleum industry reeled in shock and denial. Shell officials were furious that the talk had happened at all, and several days of closed-door boardroom meetings ensued to discuss the potential monetary backlash of Hubbert's predictions on the industry. Incidentally, the unpopular prophet was not fired from Shell but worked there until he retired in 1963, when he became a research geophysicist with the U.S. Geological Survey and Professor of Geology and Geophysics at Stanford.

Hubbert later spoke out in an interview with oral historian Ronald E. Doel, from the American Institute of Physics:

> That caused a jolt...the first reaction was honest incredulity. Then the industry split. One side refused to accept the situation and started changing the figures. The other side, people like Shell, found they could not change the figures.
>
> Well, after about a week, when the responsible people did begin to come back, I think there were some pretty red faces in the New York office and maybe even in Houston. For one thing, they had a chance to look at their data, and they found they couldn't disprove

anything I'd said. So the whole thing had been a tempest in a teapot by people who didn't know what the hell they were talking about.

By the mid-1970s, it became clear that crude oil and gas production estimates by the U.S. Geological Survey were heavily padded. When an independent appraisal group was set up to review the numbers, the new estimate for total crude oil production came down by about 15 percent from the skewed figures.

"At least this country was back on track," wrote environmental activist George Pazik, in his highly referenced peak oil editorial, published in 1976. It "ended a 15 to 20 year period in the history of the U.S. Geological Survey which can only be classified as a sorry episode." Pazik wrote in "Our Petroleum Predicament":

> Now the bubble has burst. Our big, long spree based on cheap and plentiful petroleum is over. North America never had more than 15 percent of the world's petroleum, and the U.S. share is now more than half gone. We are using petroleum faster than any nation on Earth. No amount of money is going to find oil that isn't there. We have now reached the peak from which we can clearly see the way down the slope.

Another serious wake-up to "our petroleum predicament" came in February of 2005 with a grave and urgent report written by a former senior energy advisor with the leading Fortune-500 science solutions firm Science Applications International Corporation (SAIC). "The Inevitable Peaking of

World Oil Production" by Robert L. Hirsch was commissioned by the U.S. Department of Energy, and its predictions made it no longer possible for government to ignore "an unprecedented risk management problem" waiting to happen.

The Hirsch report stated that oil peaking will be "abrupt and revolutionary," that without timely action by government "the economic, social and political costs will be unprecedented" and that it may, in fact, already be too late "to avoid considerable discomfort or worse." Hirsch warned:

> Oil is the lifeblood of modern civilization. It fuels most transportation worldwide and is a feedstock for pharmaceuticals, agriculture, plastics and a myriad of other products used in everyday life. The earth has been generous in yielding copious quantities of oil to fuel economic growth for over a century, but that period of plenty is changing.

On the bright side, he explained that the world already has several technologies available to begin mitigation efforts at once, but that the will to act quickly and assertively is key.

"Viable mitigation options exist on both the supply and demand sides, but to have substantial impact, they must be initiated more than a decade in advance of peaking," he wrote.

Some experts believe we've already reached the summit of world oil production and are now in irreversible decline, whereas others claim we have another 10 to 15 years before global demand surpasses the earth's finite supply of petroleum.

Either way, Hirsch contends we're already behind in our response because it would take 20 years of serious mitigation efforts to transition without major societal impacts and at least 10 years with only moderate consequences. If the world waits until oil production peaks before it initiates its "crash program," the world will face a "significant fuel deficit" for at least two decades.

"The world has never faced a problem like this. Without massive mitigation at least a decade before the fact, the problem will be pervasive and long lasting," concluded Hirsch. "Countries that dawdle will suffer from lost opportunities, because in every crisis, there are always opportunities for those that act decisively."

Clearly, for too long, the United States—and much of the industrialized world—has maintained blind faith in a bottomless abundance of oil.

This stalwart position is reflected in the American industry's stock public relations statement issued through the 1960s: "The United States has all the oil it will need for the foreseeable future."

In reality, this statement was far from the truth, says Hirsch, in his report. By the mid-'60s, all the largest oil fields—the so-called "super-giants" because they were "the easiest to find, the most economic to develop and the longest lived"—had already been discovered in the Middle East.

Indeed, even Dr. Sadad al-Husseini, a retired senior Saudi Aramco oil exploration executive, reportedly told media in 2007 that the world was headed for an oil shortage.

"In [al-Husseini's] words, 'a whole new Saudi Arabia [will have to be found and developed] every couple of years to satisfy current demand forecasts.' So the messages from the world's 'breadbasket of oil' are moving from confident assurances to warnings of approaching shortage," corroborated Hirsch.

By the 1970s, in the U.S. lower 48 states, which once made up the world's richest geological region, most of the oil production had peaked and started to decline.

During a 2008 interview with the Allianz Insurance Group, the energy expert discussed where we're at with the problem of declining oil production, and what we need to do to prepare for "our energy future."

Hirsch explained that new oil discoveries—such as those made recently in Brazil, in the Arctic or in Canada's oil sands—will be of value as the world revises its energy paradigm; however, they will never be enough to "turn the tide."

He noted that China has already begun working on coal liquefaction and other new energy technologies. It has also started buying into oil fields around the world to ensure bargaining power later. Similarly, as oil peaks, other oil-exporting nations are exploring wind, solar and biofuels, and will reduce oil production to stockpile their resources for the future.

"The King of Saudi Arabia recently said that any new oil fields discovered in Saudi Arabia will not be developed; they will be saved for future generations. It is the sensible thing that many of us would do if it was our responsibility to run an oil-producing country," said Hirsch in the interview.

Canada is even thinking of creating a Strategic Petroleum Reserve like that of the U.S. and is currently the only oil-producing country—and the only industrialized nation in the West—without one.

As oil peaks, supplies dwindle and demand soars, oil as a currency will go through the roof; it will be each nation for itself, and the nations with more oil to sell and use stand to gain inordinate amounts of power.

Said Hubbert, "We can buy all the oil we want to buy from our 'friends.' All we have to do is to pay their price and meet their terms. You guess if the terms will get easier and the price will get lower as our own supplies get closer to running out."

In most countries, the problem of peak oil is "still unthinkable" and politically incorrect, says Hirsch. "Most of the rest of the world is not in very good shape to deal with the problem."

As energy and mining companies comb the globe for new sources of supply, the Arctic is getting more attention than ever before—and not just because of its oil and gas, but also for its diamonds, gold, silver, zinc, iron and other minerals, which have long been frozen out of reach.

Arctic resources are believed to be relatively abundant—compared to what's left in the world, that is—but most of it is still untapped, stated a Canadian parliamentary report in October of 2008. The bulletin outlined current activity in the Far North, future potential and associated risks of exploration and drilling in a harsh northern environment.

Currently, Russia produces 80 percent of the oil and 99 percent of the gas in the Arctic. Alaska's North Slope contains 20 percent of current U.S. oil production. Norway and Denmark have tapped significant supplies within their Arctic jurisdiction. Canada, meanwhile, has already discovered 1.7 billion barrels of oil and 880 billion cubic metres of gas in Canada's North—about 25 percent and 33 percent, respectively, of Canada's exploitable resources in the area, according to Indian and Northern Affairs Canada (INAC).

Plans are in place to develop oil fields in the Mackenzie Delta-Beaufort Sea Basin and the Arctic Archipelago through several of its channels. The $16.2-billion Mackenzie Gas Project is key to such area development and includes a massive pipeline project to distribute gas from the Mackenzie Delta to Alberta and other markets. Imperial Oil and other partners, including the Aboriginal Pipeline Group (APG), run this project, but regulatory challenges and Native claims issues have stalled progress.

The report noted that exploration and drilling in more remote and offshore locations are further complicated by factors such as climate, supporting infrastructure and equipment, economic considerations and environmental risks.

For example, extreme cold, sea ice cover and high winds can make working conditions treacherous, limit access to target resources and drive up the technological costs of extraction. Likewise, it's neither cheap nor easy to secure a rig that can operate in such severe conditions; most aren't built for use in sub-zero climates. Melting permafrost on land that's been frozen for years also complicates the building of supporting infrastructure—roads, pipelines, ports, facilities, etc.—needed to move projects forward. These logistical challenges to production result in higher costs and increase the chances of an environmental mishap.

"Energy projects in the Arctic will bring economic benefits but will also affect sensitive ecosystems and the traditional lifestyles of northern peoples," government officials warned. "An oil spill would significantly damage the Arctic ecosystem."

Canada currently has in place its Arctic Waters Pollution Prevention Act and measures that restrict oil tanker travel in fragile areas, but as the ice melts in the Arctic and economic activity heats up, the likelihood of disturbance or disaster increases.

"According to a recent study by the Arctic Council, it remains to be seen whether climate change will make the Arctic more, or less, attractive for oil and gas activities," the report concluded.

Two decades after the *Exxon Valdez* super tanker accidentally dumped its oil cargo over the Alaskan coast, leaving behind "a thick pancake of shiny black" covering the calm waters of Prince William Sound, the World Wildlife Fund (WWF) says not enough has been done to improve oil spill response capabilities in the Arctic.

In 2009, on the eve of the spill's anniversary, the WWF renewed its plea to government to halt all oil and gas exploration in the Arctic until there is better technology in place to clean up oil spills.

It was March 24, 1989, the fateful day when the U.S. oil ship hit a reef in Prince William Sound along the Alaskan coast and ripped open, dumping 40 million litres of oil into the sea. Literally overnight the name *Exxon Valdez* became a contemptible household name and, in the days that followed, its ominous, black shadow deepened as the world viewed now-famous images of oil-drenched ducks and slick, black shorelines—incontrovertible evidence of one of the worst ecological disasters in history.

Early impact assessments revealed the death toll: 250,000 sea birds, 22 orca whales, nearly 3000 sea otters, 300 harbor seals and uncountable millions of fish eggs.

In its initial 2008 report, "Oil Spill Response Challenges in Arctic Waters," the WWF had warned that consequences would be dire should such a disaster occur. James Leaton, Senior Policy Advisor with WWF-UK, said:

> The Arctic offers the highest level of ecological sensitivity and the lowest level of capacity

to clean up after an accident. This combination makes it unacceptable to expose the Arctic to an unfettered scramble for oil.

The WWF report claimed that the same Arctic conditions that increase the risk of oil spills—extreme cold, moving ice floes, high winds and low visibility because of a reduced natural light—could also hinder effective emergency response to an oil spill.

At the time, the foundation called for new international regulations on shipping in the Arctic, route and zoning guidelines and stocking of mandatory oil-spill response equipment on board vessels destined for the Arctic.

A year later, the foundation wrote in its new report, "Lessons Not Learned," that although emergency response regulations had improved somewhat in Prince William Sound, overall, the petroleum industry was still not prepared to deal with another oil spill disaster.

According to Margaret Williams, WWF's Alaska program managing director:

> The *Exxon Valdez* spill has been the best-studied oil spill in history, and scientists have found that even 20 years later, the damage from the spill continues.
>
> Fishermen's livelihoods were destroyed, many wildlife and fish populations still haven't recovered and the Alaskan economy lost billions of dollars. We can't let that happen again in Alaska's productive waters.

Several other groups participated in similar studies with the aim of augmenting preparedness in the event of a future disaster. Given the inevitable increase in traffic and activity expected in the Arctic, it would have been negligent for industry and officials to ignore WWF's grim forecast.

The National Oceanic and Atmospheric Administration (NOAA), in partnership with the University of New Hampshire, collaborated on a review of oil spill emergency response standards, involving experts, government officials from Arctic nations, industry and Aboriginal communities, including the U.S. Coast Guard and the U.S. Arctic Research Commission.

They based their findings on five hypothetical emergency maritime scenarios: a cruise ship carrying 2000 guests and crew runs aground and everyone must abandon ship; a damaged ore carrier gets trapped in ice; an explosion occurs on a fixed drilling rig north of Alaska; an oil tanker and a fishing boat collide causing a large spill; and a tug boat pulling a supplies barge hits ground in an "environmentally sensitive" zone near the Bering Strait.

The Coastal Response Research Centre (CRRC), a partnership between the University of New Hampshire and the National Oceanic and Atmospheric Administration (UNH-NOAA), released a list of recommendations for Arctic emergency preparedness just prior to the highly publicized 20-year memorial of *Exxon Valdez* in the media. It called for stronger multinational co-operation and emergency response protocols; better logistical support for those responding to a disaster; updated Arctic weather and navigation data; research studying the behaviour of oil in cold water; new, improved Arctic spill response technology; and the designation of emergency docking ports for damaged vessels.

"The reduction of polar sea ice and the increasing worldwide demand for energy will likely result in a dramatic increase in the number of vessels that travel Arctic waters," said Nancy Kinner, a UNH collaborator and professor of civil and environmental engineering. "As vessel traffic increases, disaster scenarios are going to become more of a reality."

But the biggest problem when it comes to disaster preparedness is communication, Kinner told the media.

Those involved in the rescue can waste precious time wondering who's in charge, which countries will help if there's an oil spill, damaged vessel or endangered crew, or which ports are open for docking in an emergency.

Currently, the U.S., Canada and Russia have agreed on search and rescue protocol, says the CRRC, but that doesn't cover disaster cleanup operations—and other northern nations have yet to join.

The *Exxon Valdez* spill polluted 2092 kilometres of Alaskan coastline, killing marine wildlife and leaving enduring traces of the oil spill 20 years after the fact. The cleanup cost more than $2 billion and is not finished. Exxon, which is now called ExxonMobil, has paid damages exceeding $1 billion and will likely never live down its "bad rep."

Even Alaska's environment conservation chief at the time of the spill, Dennis Kelso, is still adamant that the Arctic's future could be endangered by the imminent surge in oil and gas exploration and that industry should tread gently.

"At a time when the entire set of ecosystems is under deep stress from global climate disruption, we would be well advised to go carefully when we think about extending these industrial activities," said Kelso.

The world's current dependency on a finite and dwindling supply of non-renewable fossil fuels presents us all—government, industry and average fuel-reliant citizens—with a number of unavoidable economic, commercial, political and personal challenges. These challenges include the rising cost of oil and gas; more volatile relationships between energy-producing and energy-consuming nations; the growing spectre of fuel shortages; expanding costs and technical considerations for petroleum companies drilling in less profitable and less hospitable areas—including the Arctic; a more vulnerable changing environment; northern cultural sensitivity; global warming; and the individual choices we now all face which will help make or break destructive cycles of consumption.

The verdict is still out on a number of important issues, but debate is heating up and government and industry appear to be working together to explore a number of options for helping the transition to new sustainable energies and for frontier exploration in a harsh deep-sea environment. Have we learned our lessons? Do we know enough about the Arctic and its changing climate to make our fervent quest for more oil and gas safe, cost-effective and, most importantly, not harmful to an increasingly sensitive and unpredictable part of the world?

The answers depend, to an extent, on who you ask, but it's safe to say that a world that runs on oil won't likely stop before it runs out. Hopefully, before that happens, our elected officials and industrial spirit will drive changes in the energy sector to make it more sustainable, affordable and earth friendly; after all, isn't "industry" the root word of "industrious?"

Hubbert, the unpopular prophet from Shell Oil Company who seemed to know what he was talking about, saw past our so-called "petroleum predicament" to a better future.

"When we combine limitless clean energy," he said, "with a new lifestyle [that] no longer insists that 'more is better' and that 'there is never enough,' we will be in on the dawning of a new age upon the earth."

And when you hit the bottom—of the barrel—there's only one way to go.

CHAPTER FIVE

Defending the Far North

North American folklore tells of an ominous ape-like creature known as a sasquatch or Bigfoot that lives in the icy trees of the Far North and frightens away intruders.

If one were to liken an Arctic nation to this mythic custodial beast, to be sure, it would not be Canada! From far and wide, the northern country is more like a gentle giant: laid-back, friendly, kind, tolerant, accommodating and more willing to discuss than fight over matters—even those of great national import and pride.

Spanning a total area of nearly 10 million square kilometres, Canada is the second-largest country in the world. Slightly larger than both China and the United States, the sovereign state is dwarfed only by Russia, which boasts the world's largest landmass—nearly 17 million square kilometres—and untold economic and strategic power as global warming and the race for resources open up the Far North.

Today, the formidable former Soviet state is said to be in the best position to take advantage of the new opportunities in the Arctic. Its fortuitous geographic position, large northern population—more than two million people—and aggressive financing of infrastructure to support oil and gas research and exploration in the High North have put Russia ahead of the pack.

Meanwhile, Canada is lagging behind and is noticeably "ill-prepared" to deal with emerging strategic northern issues, leading experts say. Historically, Canada has been criticized for its neglect of the north and lack of presence and military capabilities to adequately defend its frigid front in the face of possible threats.

As climate change opens up new shipping lanes for circumpolar transit—at first, in the summer months, then year-round—several nations are scrambling for a piece of the potential resource jackpot beneath the Arctic sea. A dramatic increase in northern traffic—by air and water, including below sea level—coupled with the melting of Canada's famed Northwest Passage present a number of imminent security challenges, both to Canada and the world.

The danger isn't so much the threat of a military attack or a full-blown World War III, but rather of some very real, entirely possible James Bond-like scenarios: cargo smugglers, illegal aliens and terrorist-related incursions via currently open and insufficiently manned and monitored northern sea ports. One such harbour is the newly revived Port of Churchill in the far north of Manitoba, which is poised to become a two-way revolving door for international trade.

The world-changing attacks of September 11, 2001, burnt the imprint of terror into our western psyche. They burst our bubble of perceived safety and security and brought into focus the reality of North America's vulnerability in the face of extremist violence.

But top Canadian defence expert Rob Huebert says our nation's leaders should have gotten serious about counter-terrorism as far back as 1985, when terrorists bombed Air India Flight 182, operating between Toronto and Bombay.

"Canada should have been investing in developing its capabilities back [in 1985]...it just goes to show how well we can ignore threats," he argued, in our June 2009 phone interview.

It's clear the subject strikes a deep chord with him, as he goes on to give the death count of the June 23 Air India massacre—329 people, including more than 280 Canadians and landed immigrants. Eighty-six were children. A separate bomb destined for a second Air India flight detonated, killing two employees in a Tokyo airport. His voice is solemn, his words unwavering.

"They had a memorial for the victims of 9/11 but not for Air India. [Our leaders] were ashamed to admit we were attacked," says Huebert. "We did start improving after 9/11. Cabinet became directly involved, working with advisors and agencies on security issues...but our infrastructure is still so limited."

Since the bombing, Huebert has collaborated on countless papers about Canadian foreign policy in the new age of terrorism. He is co-author of the report "To Secure a Nation: Canadian Defence and Security into the 21st Century," and his articles have appeared in many prestigious publications such as *Canadian Foreign Policy* and *The International Journal*.

A few years ago, in one of his policy reviews, Huebert painted the Canadian North as a weak link in North America's broad security chain:

> While no one is expecting an immediate attack from Inuvik [Northwest Territories] by Al Queda, potential dangers do exist in the long term. If southern borders are made more secure and the northern ones are not,

it stands to reason that the latter constitute a vulnerability.

He noted there is still no protocol in place for security screening of passengers boarding aircraft in many of Canada's smaller northern airports, and that terrorists could ultimately exploit this weakness and send weapons of mass destruction undetected to terrorist cells in the U.S. via Canada.

As he and I discuss the scenario of a terrorist attack from the north, Huebert is unequivocal in his response.

"The north is a vulnerable entry point that we can't overlook—but the most likely terrorist plot is the one we haven't thought of yet," he stresses.

That is why, in the post-9/11 world we inhabit, countries are now "refocusing on their foreign defence policies and rolling out action plans for rebuilding military rearmament programs," Huebert says.

"Where does that go? I don't know where that goes, but I am afraid it will go places that Canada may not be ready for."

His critics don't call him a "Principle Purveyor of Polar Peril" for nothing, he adds, half-jokingly.

Like Huebert, Canada's top military and defence officials are abundantly aware of the need to adapt to the changing face of global, national and northern security issues in a post 9/11 world.

Four of the country's leading northern experts, who've collaborated on a book about the history and future of the Canadian Arctic, say the country faces some critical challenges in this century—economic, political and environmental—as the nation is forced to defend its cherished sovereignty in a fragile region of intense, heated global interest.

The authors of *Arctic Front: Defending Canada in the Far North*—Ken Coates, P. Whitney Lackenbauer, William Morrison and Greg Poelzer (respectively a history professor, a leading northern security expert, a senior historian of the Canadian North and a political expert on circumpolar affairs) describe how global warming is changing the benign and mostly symbolic nature of Canada's past claims to a distant frontier whose future resource potential had yet to be realized:

> For decades, Canada relied on a simple geographic truth: the North was cold and ice-bound for most of the year, and no nation on Earth had the capacity to move quickly across the land and frozen waters to pose any real threat to southern Canada. [The] RCMP [Royal Canadian Mounted Police] provided the basics: the flag was flying [on] remote islands across the Arctic, Canadian law was nominally enforced and the country's claims to northern sovereignty seemed well protected from foreign challenges. Why spend money on a region that lacked economic importance, faced no strategic threats and had a tiny and widely scattered population that was content to be left alone?

As Canada kept its gaze south, in the opposite direction of the North Pole, its Arctic neighbour, Russia—who also views the north as its own—was already intractably focused on preparing for a lucrative oil and gas future in the north by heavily investing in Arctic research and development.

As early as 1920, the Soviets had created the Arctic and Antarctic Research Institute (AARI) to study the ice-bound polar regions. Between 1937 and 1991, the entire Arctic area was widely explored by Soviet and Russian manned drifting ice stations. During this period, some 88 polar crews established and lived in scientific settlements built upon ice flows that migrated thousands of kilometres over 29,726 drift days, according to the Woods Hole Oceanographic Institution.

Today, the AARI is even bigger and more high-tech: it carries out investigations in such diverse fields as oceanography, ice physics, geophysics, glaciology, meteorology, hydrology and ecology and studies how ice interacts with ship hulls and other engineering structures to improve their safety and performance in icy maritime conditions. The Russian Ministry of Education and Science cosponsors the AARI's bilateral fellowship program with the German government, to support the development of new crops of young scientists. For some time now, Russia's Arctic has been a frontier future-in-the-making.

Meanwhile, other Arctic nations are still busy compiling scientific data to prove to a UN committee that their northern coastlines are geological extensions of the Arctic seabed. Russia has already invested billions of dollars on deep-sea scientific research and has laid claim to 45 percent of the Arctic—an area that contains most of the proven and extractable gas reserves in the region.

Russia's first claims submission was rejected in 2001—because of insufficient evidence—by the

Commission on the Limits of the Continental Shelf, a panel of United Nations scientists charged with making decisions about expanded territorial rights in the Arctic Ocean. However, Russia has since led several more data-gathering missions and resubmitted its claims in May 2009. Canada has until 2013 and Denmark until 2014 to submit their Arctic claims under the Law of the Sea treaty. The U.S. hasn't ratified the treaty but is expected to do so in the near future. If it ratifies in 2010, it may take until 2020 to iron out its boundary issues.

Russia isn't waiting for anyone: it already operates 14 icebreakers—more than all other nations combined—in its undisputed Arctic territory and is planning for more. Its icebreaker *Kapitan Khlebnikov* has crossed the Northwest Passage a dozen times, more than any other ship, Canadian or otherwise—and for a country that claims the passage as internal waters, Canada has done little, by comparison, to create an Arctic presence.

Russia, on the other hand, is stopping at nothing to make its presence unequivocal; it plans to build 40 new ice-resistant oil platforms and 14 offshore gas platforms by 2030 and has developed partnerships with other fuel-importing nations, such as China, to help finance additional ice-resistant gas storage tankers and carriers to assist in the delivery of future stocks.

Spiegel International online news reported:

"If all goes according to plan, the first gas from the Arctic should begin flowing in 2013 or 2014," says Hervé Madeo, the deputy director of an energy consortium run by

Russia's Gazprom that's developing the Shtokman field in the Barents Sea.

Despite the financial crisis, preparations for drilling are moving forward at a fast pace. The project "has too much potential" for the global economic downturn to affect it much, Madeo claims.

Clearly, the notion that Russia is leading the race to the Arctic is a vast understatement—they're already there. As other nations try to catch up, all they can make out is Russian dust trails.

Sheer Russian might and foresight are giving other nations a run for both money and resources. How did the Russians gain the Arctic advantage and manage to get so far ahead of everyone else?

"Russia is doing this by blending some Communist-era big-spending strategies with free-enterprise principles," wrote award-winning Canadian journalist and author Ed Struzik, whose extensive research on northern issues culminate in his new book, *The Big Thaw*.

He explains that then-president Vladimir Putin knew Russia's private shipyards didn't have the capacity and backing to refurbish the nation's aging national fleet, so Putin signed a decree in 2007 that created the United Shipbuilding Corporation.

"The government-sponsored project is using the state's financial power to bolster the shipbuilding sector so that it can exploit the hydrocarbons buried beneath the seabed of the Arctic," wrote Struzik.

In August of 2008, *The Anchorage Daily News* (ADN) reported that the U.S. was losing ground in the global race to Arctic wealth because of insufficient ice-breaking capabilities. Only two of its three ships are sea-worthy, wrote ADN: the *Polar Sea* is reaching the end of its 30-year service life and the *Healy*, primarily a research vessel, can only cut through aboout 1.5 metres of ice. The *Polar Star* has been docked since 2006 waiting to be serviced. Furthermore:

> Not only is Russia's fleet larger, it's also nuclear powered and its icebreakers are bigger. The biggest, named *50 Years of Victory*, can power through more than 9 feet [2.75 metres] of solid ice without slowing down. Ice thicker than 6 feet [2 metres] reduces the strongest U.S. icebreaker, the diesel-powered *Polar Sea*, to backing up and ramming.

Canada and the United States are nowhere near prepared for the new Arctic future, says Struzik. The northern expert states that the Canadian Navy has not even used an icebreaker in the Arctic since the 1950s and currently has no capacity to navigate the icy Arctic waters. The Canadian Coast Guard's five icebreakers are old and soon to be retired.

University of Toronto law professor Ed Morgan argues that for too many years, Canada has exercised what former minister Lloyd Axworthy called "soft power" with its foreign and defence policy—or "soft thinking," as Morgan prefers to call it.

But times are changing, he says. Arctic thaw, aggressive international competition for northern resources, new and perennial sovereignty challenges and dawning security issues have the Canadian government finally shifting gears.

Still, Morgan thinks Canada has to do more to secure its Arctic sovereignty than dabble in diplomacy and respond to threats with fairly innocuous environmental protection laws—as it did in 1970 with the Waters Pollution Prevention Act, after the *U.S. Manhattan* icebreaking oil tanker traversed the Northwest Passage without asking for Canada's consent. Says Morgan:

> We enacted long-arm environmental legislation to protect the Arctic but have no means to get there to police potential polluters; we embraced peacekeeping as our primary military goal but deprived ourselves of the forces needed to fill in the space between warring parties; we endorse humanitarian intervention as a military mission but deprived ourselves of the transport aircraft needed to fly personnel and equipment to the world's inhospitable regions. The government of Canada finally appears to have turned a corner on its long lapse of focus and is now pushing a version of hard power. It's about time. If we're going to savour the international law rights of a sovereign state, we have to crack some ice.

And what better way to do so than with a new state-of-the art Arctic icebreaker?

During a late-summer visit to Inuvik in 2008, Prime Minister Stephen Harper announced that Canada would build a new $720-million icebreaker. This state-of-the-art vessel, to be named after former Conservative Prime Minister John G. Diefenbaker, will replace Canada's aging flagship, the *Louis St-Laurent*, when it's retired in 2017.

Such an announcement was a departure from the government's typical stance on Arctic issues. By 2004, Canada had only budgeted $70 million to begin seabed research in support of its claim that the Lomonosov Ridge is connected to Canada's Ellesmere Island.

A radical shift in priorities became clear during Harper's throne speech, on October 16, 2007, when he spelled out a major set of initiatives to protect and promote Canada's sovereignty in the Arctic.

"The north needs new attention," the Prime Minister said. "Defending our sovereignty in the north also demands that we maintain the capacity to act."

The new Arctic strategy unveiled by the Conservatives includes building a world-class Arctic research station; comprehensive mapping of the Arctic seabed; new Arctic patrol ships and expanded aerial surveillance to guard Canada's Far North and monitor activity near the Northwest Passage; increased military presence, including 900 Canadian Rangers to patrol Canada's northern territory; and improved living conditions for northern peoples.

That year, in addition to a new army training centre for cold-weather fighting at Resolute Bay and a deep-water servicing port at Nanisivik, on the northern tip of Baffin Island, Harper also commissioned six to eight new Polar Class 5 navy patrol ships, each equipped with a helipad and capable of slicing through ice 1 metre thick. The ship fleet has a hefty price tag—$7.4 billion—to build and maintain over its 25-year service life.

Defending the Far North

"Canada has a choice when it comes to defending our sovereignty over the Arctic. We either use it or lose it. And make no mistake, this government intends to use it. Because Canada's Arctic is central to our national identity as a northern nation. It is part of our history. And it represents the tremendous potential of our future," said the Prime Minister during a visit to the Canadian Forces Base Esquimalt on Vancouver Island on July 10, 2007.

While some law experts say Canada's bid to secure its sovereignty is being made years too late—and that Canada's claims to the Lomonosov Ridge are debatable—top military experts argue that there's no time like now to step up efforts to secure and defend Canada's Arctic, especially in the wake of aggressive Russian hubris.

When the Russians planted a 1-metre-high titanium-encased flag on the North Pole seabed in August of 2007, the symbolically charged mission drew international criticism, with Canada's Defence Minister Peter MacKay—then Canada's foreign minister—comparing the brazen act to an imperial land grab.

"This isn't the 15th century. You can't go around the world and just plant flags and say 'We're claiming this territory,'" MacKay reportedly said.

Tensions rose again, in late February of 2009, when just days before U.S. President Obama's inaugural visit to Ottawa, two long-range Russian Tupolev bombers soared by the perimeter of Canadian Arctic airspace—a dubious happenstance that MacKay sardonically blasted as a rather "strong coincidence."

Two Canadian CF-18 fighters from NORAD (North American Aerospace Defense Command) and Canada Command rapidly headed off the Russian planes, and Moscow got a terse warning from the minister to "back off" Canada's northern front. Russia responded with incredulity, calling MacKay's reaction a "farce" and iterating the flights were just a "routine" test mission. Critics suspect the fly-by was an overt probe of Canada's security-response capabilities on the eve of a major national event that commanded its full security coverage. Foreign Affairs Minister Lawrence Cannon responded forcefully to the flight controversy, stating that Canada will not be "bullied" by Russia.

Soon after the public row of words, top officials from the two countries had a private meeting, which appeared to show greater co-operation on Arctic affairs, Canwest News Service (CNS) later revealed in May. The news came along with an unnerving report that Russia's latest national security strategy would not rule out armed conflict over its share of the world's resources—including Arctic oil—within the next decade.

CNS wrote:

> The two sides appeared to be in agreement about Canada's claim to jurisdiction over the Northwest Passage and even discussed a possible joint Russian-Canadian-Danish submission to the UN to determine Arctic sea floor boundaries. But yesterday's security report suggests Russia is also bracing for more pointed conflict in the Arctic and elsewhere as it strives to secure its position as a global energy superpower.

The Kremlin paper anticipates possible struggles in the Arctic, as well as in the Middle East, the Barents Sea, the Caspian Sea and Central Asia, where outstanding disputes over maritime territory exist.

"The presence and potential escalation of armed conflicts near Russia's national borders, pending border agreements between Russia and several neighbouring nations, are the major threats to Russia's interests and border security," stated the report, which looked at possible security threats up to 2020.

"In a competition for resources, it cannot be ruled out that military force could be used to resolve emerging problems that would destroy the balance of forces near the borders of Russia and her allies."

Offering a counterbalance to alarmist headlines, *Time* magazine online wrote:

> Although it vividly outlines the worsened relations between Russia and the West, the anti-Western rhetoric is tempered with acknowledgment of the beginning of rapprochement with the Obama Administration. "Now there is a viewpoint in the Kremlin that the U.S. can be worked with," [said] Nikolai Petrov, an analyst at the Carnegie Moscow Center, an independent think tank. "Russia has come out and specifically formulated its foreign and defence policy. However, this paper is not setting out how policy will look; it is setting out the de facto situation."

While Russia's top officials have publicly stressed that they'd prefer to settle potential Arctic disputes peacefully, Russia's security report suggests that in the case of a perceived threat, military action may preclude harmonious ends.

In its new strategy paper, Moscow also condemns American plans to erect a missile defence shield in Eastern Europe, specifically the Czech Republic and Poland, which could intercept and shoot down ballistic missiles fired upon allied nations by rogue states. Washington has insisted the shield wasn't conceived to block Russia, but rather to safeguard Europe from heavily armed regimes, such as Iran or North Korea until they abandon their nuclear ambitions. Russia's security report also slams NATO's (North Atlantic Treaty Organization) encroachment on Russian borders and, in particular, its plan to offer military alliance membership to Georgia and Ukraine, two former Soviet states.

Canada's Foreign Affairs Minister Lawrence Cannon mirrored Russian candour in his response to the nation's new defence directive, asserting that the Conservative government will aim to "work peacefully" with other polar countries, but "that having been said, Canada is an Arctic power, and our government understands the potential of the north. Therefore, when and if necessary, this government will not hesitate to defend Canadian Arctic sovereignty and all of our interests in the Arctic."

Within days of the Russian report, Canadian scientists publicized their completion of a major research mission that will help bolster the country's claim to a large portion of the Arctic sea floor—and its potential oil and gas riches.

Over a six-week period, geologists and oceanographers mapped the ocean bed between Ward Hunt Island—Canada's most northerly tip—and the North Pole, with an airplane aerogravity probe and advanced sonar technology. Meanwhile, another crew tried out a robotic mapping system, borrowed from Memorial University of Newfoundland, at Alert research station off the coast of Ellesmere Island. The government is spending $4 million on two "autonomous underwater vehicles" to be built, tested and delivered by Vancouver-based International Submarine Engineering Ltd., which should be ready to scan ocean floor activity in Canada's waters by 2010.

The scientists must now study the data gathered during the spring mission and prepare Canada's official bid by 2013—to what may be millions of square kilometres of new maritime territory, said Jacob Verhoef, the federal scientist overseeing Canada's treaty submission.

So far, Norway is the only Arctic country whose claims have been successfully reviewed by the UN's Commission on the Limits of the Continental Shelf. In April of 2009, the Scandinavian nation was officially granted 235,000 square kilometres of seabed in the Arctic and Atlantic Oceans. It could take up to a decade before all treaty claims are settled.

Meanwhile, Russia is stepping up its campaign to assert ownership of the Arctic and its petroleum wealth. In autumn 2008, Russian President Dmitry Medvedev even told his country's military heads that their "first and fundamental task" is to "turn the Arctic into a resource base for Russia in the 21st century."

Less than a week after Russia raised the global spectre of an armed race to resources, the Canadian government upped its own ante—in classic Canadian form—by spinning a publicity web to impress upon Europeans that Canada is an intrinsically northern nation.

"In London, the lions of Trafalgar Square share space with the towering image of an Inuit woman and her child. In Paris, an inuksuk greets people leaving the Metro. In Oslo, Ottawa is opening an Arctic political office. And in Brussels, officials are fanning out to promote the image of a cold, northern Canada," wrote the *Globe & Mail*'s Doug Saunders on May 16, 2009, of the "aggressive" new campaign that would "brand" Canada as an "Arctic power" and as "the owner of a third of the contested land and resources of the Far North."

The message that Canadian ministers and ambassadors have been mandated to spread through channels across Europe is that "Canada owns it; hands off."

But why worry about what Europe thinks when most of its citizens live hundreds of kilometres away from the nearest Arctic coastline?

"The United States and Europe both dispute Canada's claim that the Northwest Passage is purely in Canadian territory," Saunders explains. "And Canadian officials believe that Europeans are hearing a far stronger message from Russia, which has aggressively industrialized and militarized the

Far North and claims ownership of the North Pole, than they are from Canada."

Later in May, federal officials surprisingly confirmed that Canada's spring mapping mission had actually ventured beyond the North Pole, into waters that Russia intends to claim as its own, reported The Canadian Press.

"The beaver is starting to push back against the bear in the debate over who controls the top of the world," wrote CP reporter Bob Weber.

Specifically, Canada is looking at the possibility of extending its claim to the Lomonosov Ridge beyond the North Pole.

"Although much is made of U.S. Geological Survey estimates that the equivalent of 412 billion barrels of oil lie undiscovered beneath the sea ice, jurisdiction over the pole is unlikely to bring a huge resource bonanza," wrote Weber on May 25, 2009, suggesting that Canada's efforts to extend the limits to its continental shelf is more of a political pushback than anything.

The heated debate over Arctic stewardship will likely continue to escalate as its contenders vie for bargaining rights in a world on the brink of geopolitical upheaval. Change is imminent—even necessary—but it is hard to say how events will unfold.

Russia's review of the possible military clashes over finite global oil and gas reserves was entirely realistic, says defence expert Rob Huebert. He advises that Canada take all precautions to step up its own military capacities in the Arctic while maintaining peaceful diplomacy in ongoing discussions over disputed shipping laws and outstanding boundary claims.

"The Russians have been talking very cooperatively, but they have been backing it up with an increasingly strong military set of actions," the associate director of the University of Calgary's Centre for Military and Strategic Studies told media.

"You mix uncertain boundaries with major powers and massive amounts of oil and gas, and you always get difficult international circumstances," he said. "I think the Russians have made that calculation."

Russia is not necessarily getting aggressive, says Huebert, "but is getting increasingly assertive about controlling what it sees as the future of its long-term strength."

Like a throwback to the Cold War era, nations on the other side of the fence are scrambling to prepare for the worst.

Award-winning author and journalist Ed Struzik describes in *The Big Thaw* the sober debate he witnessed in a conference room at the Canadian Forces Maritime Warfare Centre in the summer of 2008. A number of top Canadian naval and security experts were in attendance, among them, Rob Huebert from CMSS, Canadian military historian Jack Granatstein and Canadian Senator Colin Kenny, chair of the Standing Senate Committee on National Defence and Security.

That day, Senator Kenny didn't mince words when he compared Canada's Arctic policy track record to Britain's and America's hands-off approach to Nazi Germany in the 1930s. "'Our country is asleep,' he said, after describing how climate change was opening up Arctic waters to foreign ships. 'We underestimate the threats around us,'" chronicled Struzik.

In his account, the author captured the overall tone of the meeting with words by Granatstein, who hastened to put Canada's Arctic security into perspective. "We are not going to go to war over the Arctic in the future," said the respected military analyst. "But we might."

Struzik observed that Huebert may have been excused for "gloating" that day, because for close to a decade, he and Colonel Pierre Leblanc, former Canadian Forces Northern Area commander, were alone in their forecast of the imminent threats that climate change and an open Arctic would bring to North America's sovereignty and security.

Struzik wrote:

Each time a crisis arises in the Arctic, Huebert told me during a break in the discussions, the government gives the public the impression it's going to act expeditiously to assert sovereignty and security. Yet whenever it comes time to committing the resources to actually doing something meaningful about it, the government backs off in the hopes that the issue will go away.

The Canadian government has gotten away with this approach in the past, Huebert remarked, but it won't be possible in the future.

Considering Russia's latest tactics and stated ambitions, Huebert wasn't at all surprised by news that the former-Cold War nation would not shy away from a new and even colder war.

Writing in the 2009 edition of the foreign policy magazine *Great Decisions*, Struzik contemplated the possible lurking threats around the corner—the unknown, scary Bigfoot—in a cold, dark and hostile land whose sovereignty and wealth are coveted by many.

"Russia, of course, is the wild card in this risk scenario. Whether the Russians are willing to wait another decade or more to exploit what they think is rightfully theirs remains to be seen."

He argues that the U.S. must quickly ratify the UN treaty, which will give Americans some bargaining rights when it comes to assigning boundaries in disputed areas with overlapping claims.

In light of its inferior Arctic fleet, the U.S carried out initial geological research with the help of its northern neighbour. In the summer of 2008, Canada shared its *Louis St-Laurent* icebreaker with American scientists so they could begin charting the seabed in the Beaufort Sea.

Working together to address northern issues is a definite start, says Struzik, but in the coming decade, Arctic nations will need to work multilaterally to prepare for a growing number of ecological challenges and global security risks that could strike from the north.

"Canada and the U.S. were right to shrug off the Russians planting a flag at the North Pole. But what would either country do if the Russians showed up in disputed territory 5 or 10 years from now with a drilling team and a drilling platform?"

Or, more pointedly, what could they do?

And, frankly, the Russian advantage may be the least of everyone's concerns if you factor in other scenarios that could play out in a warmer Arctic with less ice, lively shipping lanes, lacklustre port security and limited emergency response capabilities. What are we ready for, really?

In recent years, the U.S. National Oceanic and Atmospheric Administration has sponsored two major forums on Arctic security, emergency preparedness and protocol. In 2007, the U.S. hired the Center for Naval Analyses to review its maritime operations, and America was forewarned that global warming could "create new havens for terrorists, trigger waves of illegal immigration and disrupt oil supplies."

"Warnings such as these fall flat because they rarely come with real-life scenarios that highlight the dangers," critiques Struzik in *The Big Thaw*.

He describes four "highly credible" scenarios, created by retired Canadian Colonel Gary Rice in 2007, which suggest the worst-case security threats and emergencies that could occur as the Arctic melts and morphs into an international hotbed of activity.

The first plot he describes involves a Chinese plane charting its course along a new Arctic route when it suddenly makes an emergency landing at Resolute Bay, a small, remote Inuit village of 200.

No one dies in the botched descent, which destroys parts of the plane, but 176 passengers are hurt and require medical attention. Only two RCMP officers and one nurse are based in Resolute, and the nearest hospital is 1600 kilometres away. While emptying the plane's cargo, a sharp-eyed attendant spots a box marked "dangerous goods" and inside discovers a 25-millilitre vial of highly contagious *Mycobacterium tuberculosis*.

In his second scenario, Colonel Rice describes a French submarine on a secret Arctic mission that runs into trouble under the ice. Severe weather conditions between Greenland and Ellesmere Island make it impossible to execute a swift rescue. Furthermore, French officials don't call NATO but decide to inform the Canadian government only.

The third scenario involves a young group of rogue Aboriginal people who are bitter over unresolved land claims and governance issues. Their people are not profiting from the lucrative gas being sucked from the region and shipped south. So they lash out by forming the First Nations Liberation Movement and blow up sections of the Mackenzie Gas Pipeline. They also block the newly finished Arctic Highway.

Colonel Rice's last scenario is an intricate Al-Qaeda and Chechen terrorist plot, originating at the Russian port of Murmansk, where a mobile nuclear device is smuggled onto the *Norsk Nova* container ship, destined for an affiliate terrorist cell in Chicago.

The container arrives at Canada's unclassified port of Churchill, which has no screening devices to detect radiation. So the dangerous cargo passes easily through security. The shipment is then loaded onto a railway car heading south, but suddenly explodes, vaporizing everything within 45 metres.

High winds tear down buildings within 150 metres, and two ships and an oil tanker are hit. Oil runs into Hudson Bay. Anyone within 1100 metres of the bomb is exposed to neutron and gamma rays that will kill them within 30 days. Those within a 5 square-kilometre radius will die, too, if they stay in the vicinity for 48 hours.

Although they seem like the extraordinary plots of current blockbuster action thrillers, these scenarios are highly plausible, say experts, given the post 9/11 world we live in today. We need to prepare for the worst—starting yesterday—and work with trusted allies to do so.

Struzik concludes:

> All four scenarios demonstrate how complicated it is going to be for one nation such as Canada or even the United States to maintain security and sovereignty in the Arctic. What Colonel Rice aptly illustrated is that complex problems require complex solutions. Patrol boats and the new icebreaker that Canada plans on building aren't going to be enough.

The four collaborating Arctic experts and authors—Coates, Lackenbauer, Morrison and Poelzer—summarize Canada's strategic shortcomings in their book, *Arctic Front*, as follows:

> Of all the polar nations, Canada has done the least to develop its Arctic potential, build a presence and protect its sovereignty; its northern region has not been integrated into the nation as a whole, nor have its distinct voice and needs been heard—or responded to—outside of local community; Canada's

science and military capacities are inadequate to exercise effective sovereignty; research funding is insufficient to properly monitor the effects of global warming; Canada's desire to have exclusive control over the Northwest Passage is unrealistic, and the country should instead focus on co-operation with allies, shared responsibility and control over the Passage.

"By any appropriate international standard, Canada's capacity to defend and protect its northern region is woefully weak," the experts write.

They do, however, commend the Government of Canada for its current efforts to improve Canada's Arctic defence capabilities and respond to threats and opportunities in the north.

They cite several good moves: the "reasonable" decision to build smaller ice-strengthened frigates rather than several costly icebreakers, showing that Canada listens to its Canadian Forces and responds rationally and realistically; plans to build a new research station and conduct full seabed mapping—albeit a decade too late; the decision to expand the Canadian Rangers and enhance surveillance capacity in the Arctic; the compromise to build one icebreaker by 2017 with a helicopter landing pad for better surveillance and reconnaissance missions; and, finally, more federal funding to assist the north in its social and economic development.

Despite excellent progress, the Canadian government has failed to set a national agenda that convinces current generations of Canadians that they are, in fact, an Arctic nation, the authors conclude.

Large numbers of Canadians have little connection to the region and feel little responsibility for its past and future. Other polar nations have a solid, modern and substantial presence in the north. If this country is going to defend its interests in the Arctic, it will have to rethink the very fundamentals of its approach to the region.

Ask the experts, though, and not everyone shares the view that Canada should defend its sovereignty through military might—even if it could. In *Great Decisions*, Struzik cites a report commissioned by the Centre for Naval Analyses, in which Retired Admiral Donald Pilling, a former vice chief of U.S. naval operations, "reiterated the long-standing view that neither Canada nor the U.S. has the military capability to handle future threats in the Northwest Passage."

Furthermore, there's the question of whether Canada even has the legal right to patrol waters not currently under exclusive Canadian control. And if not, how then will Canada demonstrate its presence and authority in the region to bolster its legal case that the Northwest Passage constitutes its internal waters?

Are "boatloads of rifle-toting Canadians" the answer to this legal Catch-22, or will an active challenge to other countries' attempts to use the passage as an international waterway stunt progress in multilateral negotiations?

"I do think that it's important to demonstrate a policing presence," Michael Byers, a professor of international law at the University of British Columbia, told the *Globe & Mail*'s Doug Saunders.

"And whether its provided by a frigate or a Coast Guard icebreaker doesn't really matter; it's having a presence and having the ability to put a handful of armed men on a vessel with a helicopter if we need to."

However, some legal experts warn that sending Canada's military into a zone that may turn out to be someone else's could provoke anger and retaliation. "They say it is inflammatory and dangerous to militarize a conflict that ought to be kept in more polite domains," writes Saunders, who spoke on the subject with Ted McDorman, a professor of coastal and marine law at the University of Victoria.

McDorman says the situation in the Arctic is pure politics. Who has the biggest icebreaker fleet or the best military capacity doesn't matter here, he explains; ultimately, the unresolved issues in the High North will be decided through the closed-door combat of boardroom politicking and diplomacy.

Says McDorman, "It's a matter of coming to an agreement between countries as to how to deal with that shelf area on Lomonosov Ridge, assuming it meets the geological requirements. Putting the military up there could just make it harder for the Russians to negotiate. It'll make them dig in a little more."

So how should Canada, and the other Arctic nations, move forward from here?

Some northern pundits—including Huebert and Lincoln Bloomfield, former director of Global Issues for the National Security Council in the United States—support the idea of an Arctic treaty that would see the polar nations working together to address common issues: security, shipping regulations, pollution control, wildlife protection, impact on Aboriginal peoples and scientific research, writes Struzik in *The Big Thaw*.

The template for such an international treaty already exists for Antarctica, but the issues and interests around the South Pole are very different because, for one, no one lives there. The Arctic is home to about four million people—nearly two million in Russia, 650,000 in Alaska, 130,000 in Canada and just over one million in Greenland, Iceland, the Scandinavian countries and Denmark's Faeroe Islands.

"The interests of these people would have to be represented and accounted for in any future treaty," writes Struzik, adding that shipping and drilling in the region are inevitable and must be monitored carefully.

It is unrealistic, he says, "to expect any country to refrain from economic activity in the Arctic given the high stakes and the investments that have been made so far."

So how the process is governed will be of paramount importance and must consider all issues around safety and the impact on northerners and the north.

Huebert has suggested that the Arctic Council, established in 1996, may the best vehicle to expedite the creation of such a polar treaty.

It already brings together eight Arctic nations who've committed to working together on the Arctic Environmental Protection Strategy. The Council also includes observing countries—France, Germany, the Netherlands, Poland, Spain and the

United Kingdom—and other organizations such as the International Union for Conservation of Nature, the International Red Cross Federation, the Northern Forum and the Nordic Council—so it's a natural launching pad for greater collaboration.

Says Huebert:

The Arctic Council was created in the hopes that it would serve as an international body to facilitate co-operation between the eight Arctic nations. At Canada's insistence, it also includes a role for the Aboriginal peoples of the north. What better time than now to use the council and the UN Convention on the Law of the Sea to resolve boundary issues and strengthen the rules governing the Arctic.

Struzik writes in *The Big Thaw* that this is the end of the Arctic we know, and that while it brings new and much-needed energy sources to southerners, it will also devastate the communities and cultures of northerners who live off the land.

"But it will also open the door to smugglers, illegal aliens and terrorists and very likely trigger a series of ecological collapses that will result in extinctions, extirpations and more species being added to the endangered list," he predicts.

Huebert wrote for *Canadian Military Journal* in 2005, "Despite [their] weak past record, there are signs that the Canadian government and the Canadian Forces are now beginning to take the security of the Arctic seriously."

What remains to be seen is whether the government will keep its resolve and spend what it needs to for ongoing surveillance and protection of the region, or whether it will downgrade efforts as public scrutiny subsides and other issues, such as the struggling economy and the looming oil crisis, steal the limelight.

"Do we become complacent as time goes on, or will we [keep security] up as we get further away from 9/11?" Huebert asks rhetorically in our June interview.

How much responsibility will we accept amid our efforts to secure Canada's sovereignty in the north, and will our plight address not only the safety and protection of people—northerner or otherwise—but also their health, culture and well-being, the changing environment and our children's future on this planet?

With eyes wide open, Canada and other Arctic nations are ready to dig in and exploit the northern frontier for its energy resources. Once all the *i*'s are dotted, and the *t*'s are crossed, let's hope the fine print of any new treaty laws does the utmost to protect us from future fallouts. As the world saw with *Exxon Valdez* and 9/11, it is difficult to predict and defend against every scenario of disaster or evil; but it is possible to anticipate, prepare and work together in the interests of security and sustainability, to be as ready as we can be.

CHAPTER SIX

A Future for Northern Peoples

On the first day of her trip to Rankin Inlet, Nunavut, Governor General Michaelle Jean skinned a freshly butchered seal, cut out its raw heart and ate it in front of hundreds of spectators.

The gesture—a brilliantly hatched media moment, replete with hardcore Jean's "when in Rome" moxie and the royal figurehead's blood-soaked hands—was a show of solidarity for Inuit hunters recently hit by Europe's planned ban on Canadian seal products.

The event also made for great headlines. A flurry of worldwide media coverage praised and criticized Jean's derring-do. Inuit leaders, northern reps and sealers' rights activists called the "hearty seal meal" a "culturally sensitive" display of support for traditional Inuit hunting practices. Meanwhile, animal rights groups denounced the seal-eating spectacle as political "grandstanding" to promote commercial sealing in Canada and fight the ban by appealing to Europeans' love of Inuit culture.

A Future for Northern Peoples

The *National Post*, at the time, interviewed a professional etiquette coach who judged Jean's active part in the remote community's feasting customs to be "proper etiquette," given her official role.

A French chef who had been serving up rare appetizers—such as seal tartar, seal pepperoni and smoked seal meat—for two years at his Montreal restaurant was bombarded by death threats after the Governor General made worldwide headlines by snacking on the marine mammal. They came via e-mail from animal-rights activists, mostly in France and Belgium, who told him he was "going to burn in hell." As a corollary, seal-appetizer sales shot up at the trendy Plateau-Mont-Royal restaurant, doubling almost immediately following Jean's famed tasting.

Mary Simon, president of the Inuit Tapiriit Kanatami organization and Canada's first ambassador for Circumpolar Affairs, applauded Jean's "northern nosh" for spreading awareness of her people's age-old customs and culture.

Simon told the media:

Hunting and eating a seal is not a political act, nor is it "bizarre" or "disgusting" as the anti-sealing lobby have commented. To us, this kind gesture is an acknowledgment by the Governor General of our culture and our dependence upon our wildlife as an important resource for our communities today.

But Jean's gesture appealed to people's emotions while deflecting important details.

The International Fund for Animal Welfare (IFAW) called out the Canadian government for sending a "misleading" message to the world about Europe's imminent embargo on seal products, which actually exempts traditional subsistence sealing from the ban and is chiefly aimed at the commercial seal hunt in Newfoundland and Quebec.

"I think it's another sad and desperate attempt by the Canadian government to blur the distinction between the seal hunts," Barbara Slee, an IFAW spokesperson, told Canwest News Service. "We do not oppose subsistence hunts."

Specifically, anti-hunt activists oppose the clubbing and shooting of seal pups, which are often just a few weeks old, and argue that the sealing industry's "barbaric" practices are endangering Canada's seal population. Those who defend the seal hunt say it's governed by strict animal welfare laws and claim it's no different from the commercial slaughter of pork or beef—except that people think seals are cute, so hunt nay-sayers anthropomorphize the mammal. Both groups, though, draw a distinction between Canada's commercial seal

A Future for Northern Peoples

hunt and Inuit subsistence sealing, a traditional way of life for many centuries.

However, sealers' rights activists see it differently. They say that although the two hunts are distinct, they are also intimately related, and fatally so. They claim a ban on East Coast commercial sealing products will destroy the market for all, and that "exemption" or not, a European ban will decimate the Inuit's economy.

After all, a little bad publicity goes a long way. Just think of how the New York fur industry was irreparably damaged by the provocative and punchy campaign with the tag line "I'd rather go naked than wear fur," led by the People for the Ethical Treatment of Animals (PETA).

When morals and money mix, the dominant message—whether real or counterfeit—can have a powerful and long-lasting effect. This is the line where politics, commerce, culture and capitalism become blurred. Ironically, it's Europe, not North America, that is home to most of the world's remaining fur farms today. Similarly, government officials celebrate the age-old Inuit lifestyle in the public eye, while politically backing industrial projects that will destroy the Inuit's native homelands and uproot their cultural traditions.

In our fast-paced culture of sound-bite politics and infotainment, too often "soft" news triumphs over the content-based, hard realities of our dirty-money world.

The European ban is expected to take effect in the spring or summer of 2010. Meanwhile, the EU's moral posturing—right or wrong—has already shaken the foundation of the Inuit seal hunt, which was once worth about $1 million a year—or 3 percent of the total hunt. Only about 70,000 seals were hunted in 2009, out of a commercial hunting quota of 273,000, according to Fisheries and Oceans Canada.

Top-quality seal pelts sold for $105 a piece in 2006. This figure sank to $30 by 2008 and to $14 in 2009, so many hunters didn't bother partaking in the seasonal hunt.

Controversy over Canada's commercial seal hunt isn't new—it has been around since the 1970s, when faux-fur and free love were all the rage. Over the years, celebrities such as idol Paul McCartney, French actress Brigitte Bardot and famous blond Pamela Anderson have spoken out against the "inhumane" slaughter and skinning of young seals used in the fashion industry. Campaigns and protests evoked heart-wrenching images of doe-eyed, white baby seals, with disturbing footage of the brutal clubbing of seal pups on blood-drenched, carcass-littered ice floes off the coasts of Newfoundland and Labrador.

The touchy subject was catapulted back into the limelight with PETA's dark, yet creative, anti-sealing parody of the 2010 Vancouver Olympics logo, depicting an Inuksuk figure with a raised club, ready to bludgeon a baby seal, with an Olympic ring dripping blood below. The over-the-top campaign was designed to speak out against MPs who had recently voted to use the upcoming Olympic Games to protest Europe's pending ban on Canadian seal products.

Even Fisheries Minister Gail Shea had backed the idea of getting Canada's Olympic athletes to wear seal on their games uniforms. "I think it's a good symbolic suggestion—to add something to the outfit of our athletes. I think it would be a good statement for the Canadian sealing industry, and Canada's support of it," said Shea.

But Canada's Olympic Committee president Mike Chambers quickly slammed the idea, because seal products on athletes' gear would not only hamper sport performance but would "politicize the games." He told the media:

> I'm used to those in the political arena wishing to attach their issues to the Olympic arena. But this is one...that will not, and cannot, be allowed to occur. It's our intent for our athletes to remain free of the politics that arises in and around the Olympic Games. The seal issue, while important, is an issue that has become politicized.

In spring of 2009, Prime Minister Stephen Harper stated that Canada would "vigorously" defend its sealing industry against Europe's ban on seal products, but not at the expense of its valuable trade relationship with the EU.

This view would help to explain the Conservatives' soft push for the symbolic seal-adorned Olympic uniforms. It also sheds light on why the Governor General may have gobbled up a seal heart for the cameras in a remote region of the Arctic that, until now, Canadian officials have shown little interest in governing, occupying or even protecting.

Were these just token acts—lip service paid to the north and a thumbing of the nose at Europe— or was there substance and fire behind Canada's sudden and passionate public defence of its Inuit people and Canada's commercial sealing industry?

As the High North becomes a hotbed of international economic interest, age-old and untimely questions around Canada's Arctic sovereignty are resurfacing.

Russia is ahead of the game—that much is clear—with strong industrial and military presence already established in the Far North. However, both the U.S. and the EU have challenged Canada's major claim to the Northwest Passage as its exclusive territory, and Canadian officials worry that Europe is getting a louder, stronger message these days from Russia, than from Canada.

So the Canadian government engaged Europe in a little conversation of its own—a friendly public relations dialogue, adjudicated by the media, to reinforce in Europeans' minds that Canada truly is a sovereign Arctic nation, and that it protects and promotes the culture and livelihood of its northern peoples by ardently defending their traditional seal hunt.

Seen in this light, Jean's new penchant for seal—which she later, reportedly, told her daughter tastes like sushi—was just a bitter bite of government propaganda, a seal sacrifice that reeked of a hidden agenda.

CBC commentator Rex Murphy's cynically responded to the Governor General's publicity stunt with his "seal of disapproval":

> The seal hunt controversy is actually the real business, not the hunt itself. It generates reams of publicity, is a great fundraiser,

and it offers politicians almost everywhere a chance to huff and puff on both sides of it to their great glory once a year, while the hunt itself declines, and the few poor sods who make an honest dollar from it are mere forsaken tokens on the board of a cynical and hollow game.

A mouthful and mind-full, in typical Rex Murphy style, but as usual, spot-on.

Global warming and the mad dash for Arctic resources do not just threaten to destroy the ecological landscape upon which the Inuit have depended for centuries. Melting ice and high stakes are grinding the wheels of politics and industry into action, and their momentum could help to make—or break—the future of a people whose resilience and cultural inheritance is being challenged by times of unprecedented change.

During her week-long trip to celebrate Nunavut's 10th anniversary as Canada's newest territory, Governor General Jean addressed hundreds of community members in a local school gymnasium. Her words evoked a brighter, more inclusive future for the north and northerners and seemed genuine and inspired, but their reception was staid and cautiously optimistic, especially among groups of more educated and politically savvy listeners.

"The reason I'm here is I really want people down south to know what life is like here," said Jean. "Development in the north cannot happen without you. It has to be about you."

A catchy slogan, to be sure, but what lies behind it? The Queen's representative has lobbied the government quite heavily to sell the idea of an Arctic university in Nunavut, where a host of social problems such as alcoholism, drug use, family violence, suicide and depression also includes the lowest high-school graduation rate in all of Canada: just 25 percent of Nunavut students earn a diploma.

Jean explains that her idea was inspired by Norway's highly successful northern University of Tromso, which has degree programs in medicine, law, geology and music and is seen as an essential step to involving Norway's Sami peoples in the development of Norway's Arctic.

Jean argues that a university in Canada's north could help prepare Inuit youth for exciting futures in diverse areas, including engineering work with mining firms stationed in the north, which would help northern people benefit from the economic boon expected in their region. Jean later told reporters in Ottawa:

> So all of Canada is now looking to the north and saying, "It's important to defend our sovereignty in the north, it's important to deal with changes from climate change, the Northwest Passage will soon be a maritime highway, it's important to explore the abundant natural resources—gas uranium, diamonds, gold."
>
> That's all very good—but at the same time we absolutely cannot forget that this sovereignty is an empty shell, the development of the north will be an empty shell, if it happens without the participation of northern people...we need to build viable, healthy, durable communities there.

It's a solid argument, and a university in Canada's North is a great idea. The problem is that the Conservatives have no intention of building such a bricks-and-mortar institution of hope for Inuit people north of 60, and Jean knows it. They have earmarked millions of dollars for northern research and training—science and military defence—and are spending another $500,000 to build up an existing network between several northern schools, including Nunavut Arctic College, which share resources and offer post-secondary studies in areas such as nursing, jewellery making and computer technology. So it looks like Ottawa's gung-ho G.G. was just spouting a whole lot of political flim-flam.

Today's political tide is changing for indigenous people of the north. Mary Simon, national president for the Inuit Tapiriit Kanatami, optimistically expressed in her buoyant and reassuring keynote address at the 2008 Arctic Indigenous Language Symposium in Tromso, Norway. During her speech, entitled "Good Intentions are Not Enough," Simon congratulated Prime Minister Stephen Harper on his "historic" apology to First Nations and Inuit people for their experiences in residential schools. She also observed soberly that it will take more than good intentions to help her people heal and move forward to build a brighter future for themselves that is culturally rooted, socially progressive, economically viable and globally connected.

Residential schools are a dark period of Canadian government education policy that saw indigenous children taken from their families and homes, and isolated from their culture and language. They were forced to communicate in the dominant language—English or French—and were forbidden to speak their native tongue.

"My own experience in northern Quebec was entering a federal school at age six, speaking only Inuktitut, and being told I would be punished if I was heard speaking my own language," recollected Simon at the conference.

Her story is all too common, shared by a generation of First Nations and Inuit people, whose sad legacy has, unfortunately, been passed on to future generations. Studies show that Inuktitut is rapidly declining as the language spoken in the home. Meanwhile, there's a growing body of research to support the idea that indigenous language education is pivotal to the health of Inuit communities.

"The greatest predictor of long-term success in school for indigenous children is how long they receive instruction through their first language," said Simon, citing a United Nations expert panel on Indigenous Children's Education and Languages.

"This is more important than any other factor, including socio-economic status, in predicting the educational success of bilingual students," she added. The UN expert panel she refers to concludes:

> Given what we know about the effects of enforced dominant language education policies that result not only in considerably poorer performance results but also higher levels of non-completion [of school], the pursuit of such policies could be said to be contrary to...the right of [indigenous children] to an education.

Simon vigorously argues that it is a human right for all children—Aboriginal or otherwise—to receive an education in their native language and that this process must not be interrupted, especially during a child's critical formative years when language is learned—as it was with the residential system.

The social and cultural fallout from this period did not just devastate the language base of a traditionally oral culture, but it also destroyed homes and spurred a dangerous cycle of addiction, mental illness, family violence, abuse and isolation that still stigmatizes Inuit and First Nations' communities today.

Here, "more than good intentions," says Simon, means that both the Canadian government and Inuit communities must work together to fund Aboriginal language programming that acknowledges the importance of the Inuktitut language to the health of Inuit communities and the basic human right of Inuit people to learn and "reclaim the legitimacy" of their mother tongue.

She adds that her people have petitioned the government for an Inuit Knowledge Centre that would focus on research conducted on Inuit culture by Inuit people that will help to develop Inuit scholars in the future.

Says Simon, "Our language is who and what we are...and the health of our language lies at the core of our well-being."

In recent decades, the Inuit people of Canada have mobilized politically. Much has changed for them, and much is still changing. Drawing on the support of bodies such as the Inuit Tapiriit

Kanatami, the Inuit Circumpolar Council and regional Inuit organizations, the Inuit have raised the profile of northern people and northern issues. In 2007, the last of four historic Arctic land claims agreements was signed, which re-establish indigenous rights to land and resources for all four Inuit regions in the north.

The road is still long, but Inuit activists and leaders, like Simon, say that as community rebuilding efforts continue, the voice of Canada's Native northerners is getting stronger. Today, Canada's Inuit have the political clout to demand the right to participate in decision making around the sustainable development of their homelands. They also have growing means and support for fostering initiatives to preserve and protect their traditional culture and practices. Finally, they have the legal right and organizational power to partake in the economic opportunities and rewards of mineral, oil and gas extraction in the north.

In its 1997 report "Canada and the Circumpolar World," the Canadian government's standing committee on Foreign Affairs and International Trade, chaired by MP Bill Graham, acknowledged:

> It is clear both that Arctic indigenous peoples want 'in' to the processes of circumpolar policy development, as a matter of right and not of grudging privilege or convenience, and that they want the Arctic states to give international recognition to this right.

A half-decade later, however, at the 2003 climate talks in Milan, Italy, the Inuit were shut out of discussions. During one of the hottest years on record, delegates from most of the world's nations met to iron out details of the Kyoto protocol—the first-ever international agreement on greenhouse gas emissions reduction—but the Inuit were not an official nation-state, so they had no voice at all.

Ironically, it's the Inuit who are the most affected by the catastrophic consequences of accelerating global warming, which is being spurred on by the modern world's libertine fossil-fuel consumption.

At the time, Sheila Watt-Cloutier, chair of the Inuit Circumpolar Conference (ICC), attended the forum and spoke to the world press on her people's behalf. She announced that the Inuit people of North America were launching a human rights case against the U.S. government for refusing to ratify Kyoto and further endangering more than 155,000 indigenous people who live in the Arctic and face extinction because of global warming. Watt-Cloutier argued:

> We are already bearing the brunt of climate change—without our snow and ice, our way of life goes. We have lived in harmony with our surroundings for millennia, but that is being taken away from us. We want to show that we are not powerless victims. These are drastic times for our people and require drastic measures.

The announcement got lots of media play but, like most stories, was blotted out by the next splashy headline.

It's debatable how much actual off-paper progress has been made by, and for, Inuit people today, critics argue. In their unapologetic portrayal of Canada's lacklustre defence of the north and northerners, the four Arctic experts—Coates,

Lackenbauer, Morrison and Poelzer—paint the Inuit as a trivialized and exploited underdog in a wild, pitiless pursuit for Arctic resource riches. In *Arctic Front,* the foursome write:

> Into the increasingly ice-free waters race government scientists, capitalists and the military, as the industrial world seeks the additional supplies of oil and gas necessary to maintain Western styles of living.
>
> Indigenous leaders, whose claims and accomplishments grabbed headlines a few years back, have been reduced to bit players—and there is an unstated recognition in government and development circles that one of the key attractions of the High Arctic is that there are virtually no indigenous people living there, and thus no one to consult before development takes place. For developers now used to adapting to the realities of indigenous autonomy and expectations for local control, the prospect of working in a largely indigenous-free zone is a dream come true.
>
> For the Inuit, it's a nightmare in the making.

The Canadian Arctic is home to about 45,000 Inuit—or less than one-third of 1 percent of the Canadian population—spread across 53 communities. Most of Canada's Inuit live in one of four regions, which are known as "Inuit Nunaat," or "Inuit homeland," in the Inuktitut language: in Nunavut, Nunavik (northern Quebec), Nunatsiavut (Labrador) and Inuvialuit (Northwest Territories). Such a small group spread across such a massive area makes the Canadian Arctic a remote and sparsely populated region.

The Inuit—which simply means "the people"—are a circumpolar race that lives not only in Canada, but also in Russia, Alaska and Greenland. They share a common culture and language and descend from the Thule people, who came from Russia to Alaska in about 500 AD and then migrated east to Canada by 1000 AD.

The 2006 census reported that Native groups make up just 15 percent of Alaska's population, around 85,000 people. Almost half of the population of the Northwest Territories is indigenous, as is a quarter of the Yukon's population. Nearby neighbour Greenland is home to about 45,000 Inuit, and approximately 60,000 Sami people live in what's known as Euro-Arctic: 40,000 in Norway, 15,000 in Sweden, 4000 in Finland and less than 2000 in Russia's part of the Kola peninsula. Russia, on the other hand, has over one million indigenous people living across its northern reach.

The Inuit live off the land in one of the harshest environments on Earth. They are adaptable and resilient, with wide knowledge of the natural world. However, they are also more vulnerable than most to the dramatic effects of global warming and the wave of industry that a milder climate is drawing northward.

Today, Inuit people straddle two worlds: one traditional, one modern. What was once familiar is becoming foreign and unpredictable. What was once remote is setting up camp in their backyard. Their ancient oral culture has always relied on its elders to transmit knowledge of the land, weather, plants and animals to its community, and on to new generations, but climate change is disrupting this cultural continuum and confounding their age-old way of life.

One Inuit resident from Puvirnituq, Nunavik, reportedly said, "There are so many changes, the older generation [is] not able to teach our children about these things anymore."

Another Inuit dweller, from Nunatsiavut, affirmed, "Weather forecasting is difficult now. Elders are not predicting the weather because they do not feel that the prediction will be reliable."

Their candid comments—invaluable first-hand observations of the major changes already occurring in the north because of global climate change—are recorded in an enlightening and personal book called *Unikkaaqatigiit: Putting the Human Face on Climate Change.*

It was written with the help of Inuit residents from 17 communities across the Canadian North and by several collaborators from the Inuit Tapiriit Kanatami of Quebec, the Nasivvik Centre for Inuit Health and Changing Environments at Laval University and the Ajunnginiq Centre at the National Aboriginal Health Organization.

In this grassroots chronicle, Inuit people share their intimate knowledge of the land, sea, sky and wildlife through direct observations and experiences with climate-related changes resulting from global warming. They discuss how they're coping with, or might deal with and adapt to, its impacts.

Warmer weather, less snow, thinner ice that melts faster in spring and increasingly unpredictable weather conditions are wreaking havoc on the traditional Inuit hunting lifestyle, they say. Animal migrations are changing. As new, exotic species appear—some of which the people don't even have names for—it disrupts the natural food chain, stressing existing populations, sometimes leading to species starvation and extinctions.

Early ice melt is also creating dangerous conditions for hunters travelling across ice in springtime. Some Inuit have fallen to their death through thinning sheets while hunting seals, caribou or polar bears. Unpredictable weather and dangerous ice make it harder for hunters to track caribou; they often travel longer distances to find food as caribou populations have begun to migrate farther north because their local habitats have been upset by increased hunting and industrial development. More rapid freeze-thaw cycles are also killing the lichen that caribou feed on, so the animals, too, must travel farther in search of food.

This tidal wave effect is not just upsetting the Inuit's sense of food security but also their cultural cornerstones. Caribou is important to the Inuit, and not just for meat; it appears on their clothing and in their art, oral legends and folklore. The animal is as much a valued cultural symbol as it is a food source. Global warming has caused caribou herds to decline in many areas. It is also endangering populations of polar bear, seal, reindeer, several bird species and a host of other marine and land mammals. The ripple effects go on.

Because travel is more dangerous and uncertain, Inuit people are travelling on land less, travel routes are often longer and fuel costs higher. With fewer opportunities to hunt, communities are seeing a decrease in hunting revenues; they have less local food available and must depend more heavily on store-bought food. This, in turn, drives up food-supply costs and has health implications, both dietary and psychological: less time on the land for many community members takes a physical, mental and emotional toll.

Unikkaaqatigiit concludes that the unique voice of northern people must be heard and given the most weight as hybrid solutions and adaptations to global warming are developed—and quickly—in conjunction with local, northern and national government officials. The authors write:

> There is a great injustice in the fact that a relatively small group of people, who have lived for thousands of years in harmony with their environment, are now being the hardest hit by the impact of development—primarily based outside their regions—on the global environment.
>
> Climate change is and has been a reality in the Canadian Arctic for a number of years. Although national and international governments have been slow to develop adaptive responses to address the impacts, Inuit do not have the luxury of being able to wait.

The Arctic Climate Impact Assessment released in 2004 made several key findings, none of them new to northerners.

It confirmed that among the cultural and economic impacts facing indigenous people, food security and human health are at great risk because of changes in species' range and availability, shrinking marine habitat for polar bears and seals, and decreased access to food sources and breeding grounds for caribou and reindeer. Also listed of great concern were elevated ultraviolet radiation levels and diseases brought by new migratory bird species to the Arctic, which can carry the mosquito-borne West Nile virus, a potentially deadly pathogen to humans.

Short of relocating all Inuit people who live in potential Arctic danger zones—a costly and mammoth process with dangerous ethnocentric echoes and implications—some major adaptations will need to be made to help northerners adapt to the effects of climate change in the north, the report concluded.

In the meantime, the Inuit people have started inventing new words for phenomena they've never seen before, such as "sunburn" or "bumblebee." Some northern fishermen now drive bigger, faster boats to travel rougher, less predictable waters and use GPS navigation systems in case a blinding storm strikes out of nowhere. Certain elders won't predict the weather anymore, and other community members advocate using enhanced communications so the community can "tune-in" to regional weather forecasts to prepare for the hunt.

It's easy to see that many of the "modern" adaptations used to reduce the impacts of global warming will also risk further estranging Inuit people from their traditional way of life and reliance on the land—the essence of what has made them who they are. How much can they endure?

Inuit activist Watt-Cloutier has famously said, "When we can no longer hunt on the sea ice and eat what we hunt, we will no longer exist as a people."

As the wheels of industry and climate change crank into full gear, these northern impacts will only intensify, she warns.

Climate change is not just an environmental issue with unwelcome economic consequences. It really is a matter of livelihood, food, individual and cultural survival. And it is absolutely a human issue affecting our children, affecting our families and certainly our communities. And the Arctic is not a wilderness or a frontier. It is our home. It is our homeland.

In a public speech, the former ICC chair called upon the global community to re-establish its connection to each other and the land—no small task, to be sure, but a major paradigm shift that would require a change to the faces of global economics, politics and social welfare systems as we know them. During her address, Watt-Cloutier poignantly asked, "Is it not because people have lost that connection between themselves and their neighbours, between their actions and the environment, that we are debating this issue of climate change in the first place?"

Hers is a weighty question, and one whose answer can't even put a bandage on the massive ecological boo-boo of our times; but it could point the way to a safer, more inclusive and sustainable future for the north, it's people—and the rest of us.

These questions are worth asking. Never has global climate change been so high on the political agendas of industrial nations. This growing social conscience, or at least the pretence of one, is as much a result of political pressure and unavoidable scientific fact as it is a growing awareness of the grave consequences we face if we continue to ignore the issues.

Prime Minister Stephen Harper travelled to the north in 2007 to promote federal plans to boost Arctic defence and spend $150 million in research funding for the International Polar Year—a massive global scientific undertaking to study the effects of Arctic global warming. Researchers would monitor, among other trends, the rises in mercury levels, as temperatures climb and multi-year ice melts, releasing pollutants that then make their way into the local food chain.

"Levels of mercury, a potent nerve poison, have long been rising in the north as emissions from sources such as coal-burning power plants in southern latitudes drift into the Arctic and remain," reported the Canadian Press, in May of 2009, following the release of a major study by the Department of Fisheries and Oceans on pollution and climate change.

Scientists now believe that climate change may be boosting levels of mercury in beluga whales, cod, Arctic seabirds and seals—one of the main traditional food sources for Inuit people. They have also found high levels of mercury in the blood and tissues of Inuit men and women. While seals are still considered a safe and healthy food for northerners, health officials responding to the study said that as ice melt continues, the situation will need to be monitored closely.

As studies like these help put a human face on the problems of global climate change, Inuit organizations are stepping up their demands that economic development in the north begin by addressing the many social and community development issues, such as health care, housing shortages, education and training for Inuit people.

Inuit Tapiriit Kanatami president Mary Simon argues that the Canadian government must work with Inuit leaders to put new northern policies in place that protect Inuit children, first and foremost, because that is the real future of the Arctic.

"The current government is giving greater profile to Arctic issues than a number of governments did in the past. That is, by itself, a good thing but only really valuable if translated into concrete action," she told the *Globe & Mail* in early 2009.

While federal spending on military development in the north is an important part of building Arctic sovereignty and security in the circumpolar world, says Simon, there is no real security without protecting "the human dimension" of northern issues.

No amount of political posturing about the culinary delights of seal meat nor the lyrical poeticisms of Speeches from the Throne on the "beauty and bounty" and "limitless potential" of our Canadian North can bridge such a distressing gap. With no real environmental or cultural security, she says, there can be no real and meaningful sovereignty.

Simon adds:

If we're not going to get our kids through school, the involvement of Inuit in economic development will continue to be slow.

Inuit now await to see whether badly needed investments on the domestic side of sovereignty enhancement will be made soon. The other critical part of any new focus on the Arctic is working in partnership with Inuit, not over our heads.

An example of such mindful stewardship is how some business has been conducted around the Mackenzie Valley pipeline project, which was conceived in the 1970s and trumpeted as "the biggest project in the history of free enterprise."

Proposed plans to build the $16.2 billion pipeline, which would transport natural gas from the Beaufort Sea through Canada's Northwest Territories to link up with pipelines in northern Alberta, were shelved following the famous Berger Inquiry.

The in-depth probe into the controversial project, conducted by Justice Thomas Berger, concluded that First Nations people did not stand to benefit in any way from the project, and, furthermore, that the ecology of their traditional lands would be heavily compromised by a traversing pipeline and its supporting infrastructure—roads, airports, maintenance bases and new towns. Berger recommended a 10-year moratorium on the project until Aboriginal land claims and conservation issues could be properly reviewed and settled.

Before arriving at his decision, Berger took his investigation to Fort Good Hope, Northwest Territories, in the summer of 1975. At the witness table, Chief Frank T'Seleie, a bright and tenacious university-educated chief in his mid-20s, was seated next to Robert Blair, the president of Foothills Pipe Lines Ltd. of Calgary, one of the pipeline's major contenders.

When it was his turn to testify, T'Seleie shocked the gallery with his forceful oratory. He said:

> You are like the Pentagon, Mr. Blair, planning the slaughter of innocent Vietnamese. Don't tell me you are not responsible. You are the 20th-century General Custer. You are coming with your troops to slaughter us and steal land that is rightfully ours. You are coming to destroy a people that have a history of 30,000 years. Why? For 20 years of gas? Are you really that insane?

The pipeline was delayed for more than a decade. In the meantime, history was made. First Nations groups in the Northwest Territories—the Inuvialuit, Gwich'in, Sahtu, Metis and others—settled land claims issues and got aggressive about planning for the economic and social futures of their communities. To this end, they formed the Aboriginal Pipeline Group.

The group's stated mandate is "to represent the interests of Aboriginal people in the Northwest Territories in maximizing the ownership and benefits in a Mackenzie Valley natural gas pipeline."

APG partnered with Shell Canada, ConocoPhillips and ExxonMobil Canada Properties and secured one-third of ownership rights to the Mackenzie Valley natural gas pipeline. By early December 2009, the project was still awaiting approval by regulators, who were reviewing the environmental and socio-economic costs of the proposed 1200-kilometre pipeline. If it goes through, Aboriginal groups for the first time in Canadian history will participate as owners in a major, multi-billion dollar industrial project.

But will it make or break them?

In 2005, following pressure from both industry and Native groups, the federal government announced it would spent $500 million over a decade to address the many socio-economic issues facing First Nations people in the north. The regulatory process has taken longer than expected. By this book's mid-December 2009 publication deadline, the report had yet to be released. The pipeline project, if approved, is projected for completion in 2014.

Meanwhile, energy companies already working in the Mackenzie Delta have become incensed by the long wait. Some, such as MGM Energy Corp., have put further drilling on hold until the decision is finalized. They were originally hoping to start moving gas by the end of 2009.

When, and if, the pipeline project proceeds, many issues will still need to be considered, some technical, some environmental and some socio-economic.

For example, building in an area of high permafrost, where the ground stays frozen year-round, could cause melting and freezing of ice as a result of heat from pipes and damage—or "heaving"—to anything built upon the permafrost, including buildings, roads, equipment and the pipelines themselves.

Building the pipeline will also cut through intact boreal forests along the Mackenzie River, damaging the habitat of woodland caribou and grizzly bears and affecting the biodiversity in the area. Some of the Mackenzie gas is slated to help further develop Alberta's oil sands, which produce some of the most harmful oil for the atmosphere.

This will increase greenhouse gas emissions and drive up production at a time when high fuel costs plague consumers and Canada should be working at meeting its Kyoto Protocol targets, environmentalists say.

The Sierra Club of Canada stated in a news release:

> Instead of subsidizing multinational oil companies to build environmentally destructive mega-projects, the federal government should focus on investing in energy efficiency and alternatives to fossil fuels.
>
> The federal government deserves credits for its commitments in energy efficiency and alternative sources of energy, but more needs to be done. Such investments are the way to meet Canada's Kyoto Protocol commitments as well as provide relief to consumers and small businesses—in northern and southern Canada—from higher fuel prices.

Also, while the pipeline project will create upwards of 2600 jobs during the construction phase for locals and temporary workers shipped into the region, the end of construction will mean regional workers will have to look for work elsewhere. So the local labour benefit will likely be short term, while the long-term social and cultural fallouts could be unremitting.

For the time being, however, it appears that First Nations groups in the Northwest Territories have taken an "if you can't beat 'em, join 'em" attitude. But is their buy-in a sellout or a hand-up from their current troubles as an impoverished minority group?

It remains to be seen whether the economic rewards of this historic partnership will be enough to offset the socio-economic and environmental costs of blending oil and water—modern capitalist ambitions and a traditional way of life that is already endangered.

Whatever global climate change brings, the future of Inuit people must be a national concern. Their existence must be protected by creative and concrete government policies and empowered by the Inuit's own sovereignty rights as Canadians and northerners.

Development in the Arctic is a future certainty; what is unknown is how this new chapter in First Nations and Inuit history will be written, and by whom. May the legacy that's passed on to future generations tell a different story from the past and speak of Canada as a truly sovereign Arctic nation with genuine respect and heart for its people.

CHAPTER SEVEN

Rolling with Climate Change: Greener Industry, Sovereign Responsibility

Today's daily six o'clock news experience feels incomplete without the latest up-to-the-minute bio-horror story on how global warming is destroying our planet.

These typically negative, doomsday threads spin a web of fear and armchair inertia around viewers, as fact-laden experts confirm that the damage has already been done—the wheels of modernity are no longer an ecological train wreck waiting to happen but rather a full-out derailment in progress.

Higher temperatures and receding glaciers are causing severe droughts in some of the world's poorest, most vulnerable places, and they are expected to worsen with time. Climatologists forecast more natural disasters as a result of severe, unpredictable weather affecting many coastal regions around the globe, as we've already seen with catastrophes such as the 1997–98 El Nino, the '04 tsunami and Hurricane Katrina.

The spectres of new migrant diseases, higher food and gas prices, freshwater shortages, increased famine and destabilizing shifts in global ecosystems are now common parlance among scientists, economists, politicians and Discovery Channel buffs alike.

The world is waking up to the reality that climate change, brought on primarily by the modern world's excess burning of fossil fuels, is gaining dangerous momentum. We can no longer sit back and ignore the mess we've made, like it isn't ours, by right, to clean up.

As His Holiness, the Dalai Lama—a devout Earth patriot and protector—once said at a news conference, "Our beautiful world is facing many crises....It is not the time to pretend everything's good."

Likewise, the subversive rhyming kids' author and cartoonist Dr. Seuss makes a similar point in *The Lorax*, a children's fable about industrialized society's impact on the environment: "Unless someone like you cares a whole awful lot, Nothing is going to get better. It's not."

Today we're bombarded by millions of sound bites and images depicting the many actual and imagined realities of global warming. It has begun to creep into our collective consciousness, upsetting our sense of security and our status quo, and, in some cases, challenging how we choose to live, work and play.

These days most people are hip to turns-of-phrase such as "green living," "carbon footprint" and "eco-terrorism"—even if they evoke but a vague sense of recognition, personal accountability, unfocused guilt or fear that we have bitten the hand that feeds us and are now at the mercy of unknown forces, omnipotent, hostile or otherwise.

Our welfare and safety have been compromised, and it's happened by our own doing—through the cumulative choices and short-term vision of our bottom-line governments, industries and consumer citizenry.

To be sure, climate change and its consequences are not always a bad thing; some of its variable side effects may be positive or even downright favourable to the livelihoods of certain communities and ecosystems.

For example, warmer weather in some parts of the world could extend the growing season, boost farming productivity and lower winter mortality caused by influenza and other cold-weather ailments. An ice-free Arctic is already opening up previously remote regions to new travel, tourism and more commercial and shipping activity. This includes lucrative oil and gas ventures, some of which will help northern communities develop and thrive. Today there is a wide-open market for new, eco-friendly and energy-efficient products and services, so depending on who you are, opportunity in the Arctic now is unprecedented. What remains to be seen is what we will make of it.

Canada's and other nations' concurrent ambitions to lay claim to parts of the Arctic and secure their northern sovereignty is no small feat. After all, having authentic power, true authority, within a territory comes with great responsibility that goes beyond safeguarding borders and people.

As aspiring governors and caretakers of the Far North, these nations inherit a moral burden along with their sovereignty status. We must not only make and enforce laws that preserve territorial integrity and protect the health and livelihoods of inhabitants. It behooves us all, to use the environmental lessons of the past to build a future that is greener, wiser and more sustainable.

After all, if we can't offer protection to our unique northern landscape and promote industrial practices and consumer choices that reduce waste, save hard-earned money and preserve the earth and our connection to it, then our sovereignty is really a moot point.

This intrinsic duty applies, on a wider scale, to all nations looking to move forward rather than backward in these challenging times and build a brighter world for their people that is safe and ecologically sound.

Today, in Canadian media and politics, most talk of climate change focuses on greenhouse gas emissions. Soaring atmospheric carbon dioxide levels are serious business, but there are other significant causes and corollaries of global warming that greatly impact our world.

We're not just running out of oil. We're running out of water! Unsafe, contaminated drinking water is a major problem worldwide, and freshwater shortages now affect 80 countries around the globe. The United Nations predicts almost half of the world's population will face severe water shortages by 2030. Pollution, drought, over-consumption and mass diversion of freshwater supplies to thirsty, growing economies are rapidly draining a non-renewable resource that we depend upon for nothing short of life itself.

It kind of makes our peak oil crisis pale by comparison.

Rolling with Climate Change: Greener Industry, Sovereign Responsibility

As our leaders in government and industry take on the task of navigating these concurrent and complex waves of change, leading us to safer shores, it will be interesting to see what creative solutions and compromises are born out of necessity.

The supply and demand economy may have steered us through the modern industrial age; and, no doubt, it will continue to separate the wheat from the chaff in the future. But, at least for a time—and in spite of the initial high costs—we are going to have to switch gears and learn to do things differently.

Because if we don't, the cost, in the end, will be too great. The climate up north, and around the world, has never been more primed and ready for change. "Till now man has been up against Nature; from now on, he will be up against his own nature," wrote Nobel Peace Prize physicist Dennis Gabor, in his book, *Inventing the Future*. He also cleverly said: "The best way to predict the future is to invent it."

Now is the perfect time to look anew at the global drama unfolding and reinvent the wheels of industry and culture to drive a more sustainable future. Even if current affairs and tree hugging aren't your passions or preferred pastimes, it's pretty hard to ignore what's taking place. Today you don't have to be an expert to realize that we've turned a corner.

We've reached a critical point where the effects of climate change can be seen and felt daily—whether filtered through the news or firsthand in your own backyard.

Consequently, there's a growing awareness of the need to embrace a greener vision—on the road, in the field and at home. More people are making a daily commitment to protect the earth by making smarter consumer choices, living more simply and naturally, and speaking up to petition their governments to revise harmful environmental policies and practices.

Even at a local level, every bit counts. In fact, it is here that change begins, takes root and grows. Today's savvy modern shoppers can buy organic products in 100 percent recycled packaging, cart their groceries home in reusable enviro-bags and slash their trash volume by disposing of scraps and leftovers in city-sponsored home composting bins. Community garden projects are taking off, and downtown city markets that support local growers are a trendy destination for Starbucks-toting Sunday shoppers.

Rolling with Climate Change: Greener Industry, Sovereign Responsibility

People everywhere are powering down to save gas, reduce pollution and slash their heating and electricity bills. They're capturing rain-water, installing sun roofs, biking to work, shifting to more fuel-efficient vehicles, exploring the benefits of wind, solar, wave and nuclear power, capitalizing on state-sponsored energy-savings rebates, and so on.

In a modest but not insignificant way, each enlightened buyer's choice, government incentive and cleaner energy solution by industry is driving a "slow" revolution. Step by step, we are learning to tread the earth more lightly. Step by step, we are moving towards a more conscious and ecologically responsible consumer culture and world. Step by step, together, we are earning our sovereign wings.

Oil and water. We say they don't mix. But arguably two of the most important and finite raw materials on the earth are united by a common reality. Both are increasingly scarce resources upon which the modern world continually depends more and more. Both are tied to a historical record of waste, pollution and over-consumption. Both are fruit of the earth, bottled or barrelled, and sold for profit without thought of running out.

Remember when plastic water bottles were eco-chic? Canadians consume nearly a billion bottles every year, making water one our country's most popular drinks. Not bad; after all, water's free of calories, sugar, sodium and alcohol, and it literally keeps us alive. The bottles it comes in, however, are another story: they require fossil fuels to produce and transport, and they eventually end up in city landfills. Then it takes 700 years for them to start decomposing, not to mention bottled water costs 3000 times more than your city tap water, which is perfectly clean and safe to drink.

But the word is getting out. Already 27 Canadian municipalities—including Toronto, London and Vancouver—have banned or limited the sale and distribution of bottled water in city-run facilities and have returned to promoting the use of publicly delivered tap water. The "back to the tap" trend has caught on at 21 Canadian universities and colleges, which now have bottle-free zones. The U.S., Europe and Australia are also kicking their plastic bottle habit.

The debate over bottled water still ignores the elephant in the room—an emerging world water crisis; but it's a start. People are starting to take water less for granted. Hopefully, the next step will be to figure out how to conserve—and share—what's left of it.

Tony Clarke of the Polaris Institute, an Ottawa think-tank that has reported on the environmental, health and economic impacts of the bottled water industry, said:

> It's becoming clear that the recent love affair with bottled water has reached its limits. Bottled water's 15 minutes are up, the marketing scam is out of the closet and the tap is back. The simple fact is that there is no "green" solution to bottled water. While it might serve a function during natural disasters or other emergencies, it is no alternative to the tap.

About one-third of Canadian households still buy bottled water, so a shift in behaviour may come slowly. But perhaps a slow shift is what we all need, says Canadian journalist and author Carl Honoré, to gear down from today's "cult of speed" to a more balanced and ecologically mindful lifestyle. After all, isn't it the mad, mindless dash toward industrial wealth and plenty—the so-called good life—that got us into this mess to begin with?

Honoré wrote, in his international best seller, *In Praise of Slowness*:

> The case against speed starts with the economy. Modern capitalism generates extraordinary wealth but at the cost of devouring natural resources faster than Mother Nature can replace them. Then there is the human cost of turbo-capitalism. These days, we exist to serve the economy rather than the other way around. Long hours on the job are making us unproductive, error-prone, unhappy and ill.

And what time and energy are left, he argues, are often devoured in manic pursuit of instant pleasures and high-cost conveniences that take a toll on our health and our environment. A slow revolution is brewing around the world, and Honoré claims this karmic coup will restructure the world as we know it. But whether change is coming too late or just in time to save the planet and its people remains a matter of perspective.

As petroleum companies continue to dig for dwindling supplies of oil and commercial developers continue to buy up remaining pockets of space in the ever-expanding urban jungle, sub-dividing them to multiply their profits, droves of city dwellers and suburbanites are staging their own quiet rebellions by getting "back to nature."

Meanwhile, nature—in the Arctic and around the world—has defied its own decree and is answering to the unnatural laws of industrial capitalism. It's becoming abundantly clear that we need a new governing mandate.

In their lives, regular citizens, like you and me, are taking up the torch to rewrite the rules of happiness and plenty. Some are renouncing their purchasing power while others are using it as green currency. Some of this change, of course, is token and superficial; so many people today define themselves by what they own, not how they live. Life is about buying, not being. Happiness and success are a measure of one's assets, not one's inner peace and prosperity.

When you identify so strongly as a consumer, says Honoré, it can be hard to kick the habit of excess consumption. Consuming less, after all, is paramount to destroying one's sense of self; going shopping, on the other hand, vivifies the materialist's spirit. Hitting the mall under the pretence of making "greener" choices, then, is sometimes the best activism we can muster. But too often we get it wrong.

The advertising industry doesn't help either, when it comes to sifting out the fraudulent eco-products from the authentically "green" ones.

The environmental movement has yielded crops of inspiration for advertising's ever-churning idea mill. In a world where image is everything, it was only a matter of time before someone made a buck off of the emerging debate around world water shortages, in particular, the pure, pristine—and endangered—icebergs of the Far North.

Remember the old 7 Up commercials? In the same vein that they sold the titillating experience of washing away dust, doldrums and thirst in an effervescent summer shower of wet t-shirts, bare skin and rain dancing, Canada's Iceberg Vodka Corporation has tapped the "mystique" and "purity" of Arctic icebergs in a bottle of booze. Yes, a bottle of booze.

The successful liquor outfit, which is run by the descendant of a long line of east coast mariners, has been harvesting icebergs for over three decades.

Owner Edward Kean spends upwards of 16 hours a day wrangling 5-tonne chunks of ice so he can then melt and siphon the clear, refreshing water of disappearing icebergs into his popular vodka. The bottle containing this premium spirit depicts a towering blue iceberg on its eye-catching label, which has an appealing and distinctly Canadian feel. As with most pop icons and images of our time, however, we celebrate the symbol at face value and ignore the deeper meaning behind it.

Sober diatribes about the world's troubles don't sell booze. It's safe to say that those who imbibe in Iceberg Vodka's purity and mystique will not likely stop to ponder the global water crisis down the pike—much less the irony of pouring the world's last freshwater icebergs into a dehydrating alcoholic beverage that the planet's most thirsty, parched nations will never drink.

We live in paradoxical times, to be sure, but some people, like European mountaineer Tim Newall-Watson, are using their entrepreneurial spirit and global conscience to help address today's major quandaries, such as worldwide freshwater shortages, by tapping the deep wells of human creativity.

Hailed as an eccentric, the U.K. native, with his team of six, drummed up the idea of navigating wind-powered icebergs across the Arctic to the Mediterranean to supply much-needed water for agriculture and consumption.

"I know we can do it; it is just working out how," he told *The Guardian* in 2004, when the project was in its early concept stage. "We just have to prove the principle that we can drag a piece of ice through the sea at a speed it would not normally go at and in a direction it would not normally go."

Newall-Watson and his team planned to attach a large kite to a mansion-sized iceberg and tow it from Newfoundland. They had already conducted lengthy negotiations with "traction-kiting" companies from Germany and New Zealand that use industrial kites to tow freighters around the world. All they needed was sponsorship of nearly half a million dollars for their test sail.

Long-range iceberg towing has never been done before, but short-range towing is routinely used in the oil and gas industry to divert huge masses of ice that have drifted dangerously close to drilling platforms. To accomplish this, a vessel navigates around an iceberg and attaches a floating cable line, lasso-style, then pulls it along a new course at a careful tension to avoid rope slippage or "rolling" the berg.

The idea of long-range iceberg towing—minus the kites—has been tossed about since the mid-

1970s, when several longstanding proposals were made to tow icebergs on "iceberg trains" powered by electric propellers from Antarctica to coastal U.S. states, including California, to provide much-needed freshwater supplies. As the problems of water shortages and climate change's negative impacts intensify, iceberg towing is increasingly being explored as a viable option to divert freshwater supplies to impoverished regions.

The first recorded instance of towing icebergs was a shorter voyage in the mid-1950s from southern Chile to the northern port city of Valparaiso as part of a refrigerated ice supply business, but it proved to be expensive, and most of the ice melted during the journey. Since then, research has advanced, and today groups such as a European consortium—made up of the Danish Technical University, two French laboratories, the Scottish Association for Marine Science, Cadiz University and Cambridge University—are working with ocean physicists and marine scientists to develop technology to help reduce iceberg erosion and melting.

The group is very interested in conducting its own iceberg-harvesting experiments and hopes to one day alleviate world water shortages by shipping freshwater icebergs to arid parts of southern Europe and North Africa.

Another out-of-the-box idea being explored by several nations is the transport of bulk water by sea in giant, floating polyurethane bags. The Medusa Bag was the first prototype, designed by James Cran of the Canadian-based Medusa Corporation in 1988, when the world became alerted to the reality of water shortages in California, Israel,

Jordan and Palestine. Early Medusa bags could carry 1000 cubic metres of water at a time, but the largest bags today are now longer than seven football fields and hold 1.5 billion litres of water, capable of transporting life-saving supplies over great distances to millions of people daily.

A Norwegian water supply company that is capitalizing on growing demand for clean drinking water has already used the flexible floating bags to ship potable water from Turkey to Cyprus. Greece has supplied bagged water to its island resorts to service their tourist industry. The buoyant sacs were also deployed during the Gulf War in support of military operations.

One day, under its New Greenland Bag Water Plan, Denmark envisions providing daily shipments of potable water to African nations in a train of five Medusa bags delivering 189.3 billion litres of water annually to countries where millions are dying from water-related diseases each year.

According to the United Nations Human Development Report, "an American taking a five-minute shower uses more water than the typical person living in a developing country slum uses in a whole day." Something to consider, while you're basking in that 45-minute shower, leaving the tap running or flushing the toilet 10 times a day.

The world's freshwater supplies are undoubtedly shrinking. The senior water adviser with the World Bank warned us back in 2000 that we're headed for a worldwide water shortage. Considering that the human body is made up of 65 percent water and that currently 40 percent of the world—more than 2 billion people—have no access to clean water or sanitation, we might just want to pay attention.

"Unless people learn to use water more efficiently, there won't be enough freshwater to sustain the earth's population," said the World Bank's water boy, John Briscoe.

His urgent concerns were echoed by United Nations Environment Program head Klaus Toepfer, who reportedly said, "My fear is that we're headed for a period of water wars between nations."

Countries at the highest risk of shortfalls in the next 25 years include Ethiopia, India, Kenya, Nigeria and Peru. Meanwhile, The United Nations Educational, Scientific and Cultural Organization predicts that average global water supplies will drop by one-third over the next two decades, and the CIA forecasts that nearly half of the world's population will inhabit "water-stressed" regions by 2015.

North America has its own share of water problems. Canadian and U.S. officials are currently collaborating to "modernize" a 1972 agreement that protects the Great Lakes, a unique system of freshwater lakes and tributaries that straddles the two nations' borders. Climate change is negatively affecting the region because of "increased population and urbanization, land use practices, invasive species [and] new chemicals..." reported the online Environment News Service.

As global warming creates dryer conditions in the southern and western states and the population continues to swell, rampant development and unsustainable agriculture will further intensify water shortages, experts say. The water is not running out literally; there are just more of us to share a finite resource whose natural abundance is at the mercy of dynamic environmental shifts beyond our control.

"The world cannot increase its supply of fresh water; all it can do is change the way it uses it," wrote the BBC in its in-depth report on water in 2003. "If we do not learn to live within our aqueous means, we shall go hungry as well as thirsty."

This is no small statement. Even doomsday prophets would be hard-pressed to come up with a more chilling and poignant argument. The BBC goes on to explain why the world's taps are running dry, concluding with the following insight:

> A world where consumption was a means to survival, not an economic end in itself, would have enough water to go round. And polluted, inadequate water might kill its children a little more slowly.

How much faith can we really place in a commercial system that routinely robs Peter to pay Paul? History tells us not much. Since the early 1900s, the U.S. has diverted its mighty Colorado River into California to meet its enormous freshwater needs; as a result, by the time the river reaches the Mexican border, it is but a trickle of its former self.

Now, parched and thirsty states in the southern and western U.S. want to divert water from the Great Lakes—a move that, if permitted, would likely wreak havoc on Canada's fisheries, hydroelectric power systems in Quebec and New York and local aquatic and land species.

The crux of the problem is that urbanization, bulging population-dense areas and commercial agriculture place massive demands on regional water supplies, which simply can't keep up. Water is even used in the oil production industry. Today, oil sands projects underway in places such as Alberta rely heavily on water. Most of the water used to extract the oil from the sand ends up as toxic wastewater. Water experts predict that climate change, combined with future water shortages, could dry up oil sands expansion.

Indeed, water and oil are mixing to create the necessary—and uncomfortable—circumstances for change.

In its 2004 exposé on selling Canada's water, the CBC wrote:

> It has been said that water will be "the oil of the 21st century," or "liquid gold," and that it will cause wars between nations. Whatever happens with regard to global water, and the environmental, economic and political fallout, Canada will be a major player.

In recent years, Canada's water debate has centred around the idea of capitalizing on the country's rich supply of water—which is estimated at between 6 and 20 percent of the world's supply—and selling it to other countries, as it does oil, gas and wood.

However, this matter is not so cut and dried. "There is also the issue of whether, under the terms of the General Agreement on Tariffs and Trade and the North American Free Trade Agreement, water is a 'vital resource' like the air we breathe, or a 'commodity,' to be sold or traded," the CBC wrote.

Time will tell whether we see fit to sell our water or share the wealth with those who need it most. However, if these so-called "water wars" turn out to be anything like our current wars over oil and gas, then, once again, the world's most powerful nations will find themselves in the privileged position to choose between old, harmful ways and new, life-sustaining ones.

At this crossroads, government and industry must decide whether the looming problem of water shortages is global or local in nature, and to what extent we have a moral obligation to protect the planet and its people.

If visionaries and entrepreneurs like Kean and his harvesting crew can get their "berg in the bag" and over to those whose rivers are toxic and dry, the water problem just might be wrestled under control.

After all, it's estimated the amount of iceberg water that dissolves into the sea annually is 3 trillion cubic metres—very close to the world's annual consumption of fresh water, which is estimated at 3.3 trillion cubic metres. In other words, there's enough for both Peter and Paul, if we just decide to use it wisely.

The alternatives aren't so appealing. On a practical level, already more than 3500 desalination plants exist around the globe, which extract sea salt and impurities from ocean water. Demands for irrigation and fresh drinking water are expected to go up in coming years, and over-pumping of dwindling supplies is exacerbating the problem. Experts say that iceberg harvesting and freshwater transport may be a better solution, given that the desalination process is both expensive and inaccessible to poorer, less industrialized nations that bear the brunt of water shortages.

A U.S. federal agency is currently conducting advanced research on how to "reclaim" water to replenish and conserve the world's depleting freshwater supplies. Worldwide awareness of the issue is growing, and market support will hopefully soon follow. So "iceberg cowboys" like Newall-Watson and his kite-wielding team, and the innovative European consortium, still might get their chance to erase thirst and change the world one iceberg at a time.

On a philosophical level, the decisions we need to make won't be easy. Those of us who live in developed nations will need to channel our energy and focus away from greed and self-interest and toward Earth-friendly notions of collective prosperity and well-being. It is a major paradigm shift, to be sure—one that challenges our very definitions of success and progress; but if we don't make it, neither will our beautiful blue-and-green planet.

The world as we know it is changing. Whether you inhabit drought-ravaged southern climes or frigid northern frontiers, lead a conventional modern consumer lifestyle or ebb and flow to the time-honoured tide of Mother Nature, your status quo—the set of expectations you've come to rely upon and live by—will soon no longer reflect our global reality.

This change may come to you in the form of thinning ice under your hunting moccasins, the plunging resale value of your gas-guzzling SUV, hard-to-swallow premiums on California grapes or the intrinsic value of fresh, potable water in warming zones of less rainfall and total glacier retreat. Water shortages are just one piece of this complex puzzle.

No one knows how world climate change is going to pan out. We can only respond to its effects and plan for the worst with best intentions. After all, it was our human ingenuity and industriousness that led to environmental collapse and critical shortage, so surely we can put our heads together to manifest a new destiny.

Around the world, there are already plenty of examples of positive, innovative changes and spin-offs in response to global warming: solar parks and wind farms, wave power plants and nuclear waste and carbon-capture storage and cleanup. Apparently, there is nothing like a peak oil crisis to fuel people's desire to invent new, creative ways to keep the wheels of modernity rolling.

Looking on the bright side, the U.K. First Post's reporter Sean Thomas mused in 2007, "Will global warming ultimately be good or bad? The truthful answer is no one knows. But 'the end of the world is nigh' makes a much better headline."

As the globe's dwindling supplies of oil and natural gas drive up electricity and fuel costs, and the Kyoto Protocol's greenhouse-gas reduction mandate takes hold, the world is investing more heavily in energy alternatives.

These "renewable" energy sources have little or no undesired consequences, unlike dirty fossil fuels and radioactive waste-generating nuclear energy. Wind is zero-emissions energy and is currently the fastest-growing energy source in the world, followed by solar and biomass, which uses organic matter such as wood or plants to produce heat, electricity and fuel for cars that are dramatically cleaner than oil.

Remember the old, picturesque windmills that dotted the countrysides of Europe and America, first to grind grain, then to pump water and generate electricity?

They're making a strong comeback, this time as high-efficiency, aerodynamic turbines made of wood and steel. The modern "wind farm" provides a "free," clean and inexhaustible energy alternative to the pollution of coal-burning factories and aging nuclear plants, which increasingly face safety issues.

Today, Ontario is the leading wind-generator in Canada. In 2008, wind accounted for 2 percent of the province's power portfolio with a supply capacity of less than 1000 megawatts (MW). The government estimates, however, that wind turbine capacity will more than quadruple by 2015, and a decade later it will account for a major portion of the 15,700 MW of power coming from renewable energy sources.

In December 2008, Canada became the 12th country in the world to surpass 2000 MW of installed wind energy capacity, with 85 wind farms operating and producing enough energy to power 671,000 homes.

Ontario was in the lead with a wind energy capacity of 781 MW, followed by Quebec at 531 MW, Alberta at 524 MW, Saskatchewan at 171 MW, Manitoba at 103 MW, Prince Edward Island at 72 MW and Nova Scotia at 61 MW.

Germany led the world with an installed energy capacity of 23,300 MW; the U.S. trailed closely at 20,413, followed by Spain at 15,900 MW, China at 9000 MW and India at 8757 MW.

Over the past 10 years, worldwide wind energy capacity has grown by 32 percent, and the Canadian Wind Energy Association estimates that by 2020, close to $1 trillion U.S. will be spent globally to increase world supply capacities well over 500,000 MW.

However, it's not all smooth sailing with wind power. The industry is facing some strong opposition from homeowners who complain that building giant turbines in their rural backyards affects property values, human health and bird populations and creates noise pollution while destroying a picturesque view. A number of wind electricity projects have been cancelled or delayed as a result, leading Ontario's energy minister to warn that "not-in-my-back-yard" (NIMBY) attitudes could threaten the province's energy security.

An exasperated Dwight Duncan told CanWest News Service in 2006:

> It's no longer NIMBY. For some people it's NOPE, not on planet Earth, it's BANANA, build absolutely nothing anywhere. None of us wants to have the transmission lines or the wires or generation sources close to home. [But] people have got to start asking themselves some pretty serious questions...

Duncan took sides with popular environmentalist David Suzuki, who created a stir back in 2005 by stating that wind turbines, if thoughtfully placed, can be beautiful, and that a "blanket 'not in my backyard approach' [to these projects] is hypocritical and counterproductive."

"[Wind turbines] produce greener power, cleaner power and we all have to do our bit..." concluded Duncan. "They are beautiful [and] of all the power sources, they are probably the least offensive from an environmental perspective."

Ahead of wind, solar energy is probably the most well-known renewable energy source in use today. Once regarded as futuristic, solar power uses cells made from silicon and other materials to capture energy from the sun and turn it into electricity. This renewable energy source also used to be expensive, but technology has come along to make solar more cost-effective and efficient. The downside of solar is that it only generates energy during daylight hours, and it can be blocked by clouds and pollution, so storage cells are required to make the source reliable and convenient.

More North American homeowners are installing solar panels to lower their electricity bills. In California, residential installs shot up 53 percent in 2004, helped by a government energy rebate.

The sun is also rising on Canada's First Light solar park, which will see a 200,000-panel solar farm constructed near the township of Stone Mills, Ontario, 30 kilometres west of Kingston. The finished 300-acre park will generate 19 MW, or enough energy to supply 2000 homes annually. Another large-scale project, being led by OptiSolar Farms Canada, is in the works in the southern port town of Sarnia, Ontario. When completed, this project will produce 50 MW of energy per year and will be the largest solar energy farm in North America.

Looking around the globe, innovative solar solutions to the world's energy problems abound. Germany's Desertec Foundation is currently working on the world's largest solar project—a 100 gigawatt (GW) solar power plant to be installed in the Sahara Desert of North Africa by 2025. The price tag on this mammoth enterprise is close to $800 billion U.S., but advocates say that using just 0.3 percent of the Sahara's lands for the plant would produce enough solar power to supply all of Europe with clean, renewable energy, meeting about 15 percent of its total energy demand. Once operational, the power generated by the plant would be sent via high-voltage direct current lines across the Mediterranean Sea and into Europe. Major potential investors who've shown interest include the Deutsche Bank, energy companies Siemans, RWE and E.on, and German insurer Munich Re.

Tidal energy is another alternative source that's getting lots of attention. The technology harnesses the power of ocean waves and converts it into electricity. Wave energy is thought to be one of the most environmentally benign ways to generate electricity. Advocates say it also has the potential to one day surpass wind and solar power because wave energy is more predictable, which increases its future chances of being delivered through a grid system and earning shareholders stable premiums. But investors need to commit now, says an American legal expert in renewable energy, or they could miss the high tide on this potential green bonanza.

As governments put pressure on the industrial sector to adopt cleaner, renewable energy standards, the public and private investors will likely turn en masse to more proven sources, such as wind and solar, lawyer Ed Feo with the firm Milbank, Tweed, Hadley & McCloy told the *Wall Street Journal* in June of 2009.

"The window for true innovation will only be open for so long...and then the demand for renewable energy will be filled by existing technologies," said Feo, head of renewable energy practices at the company's Washington office. "I hope the window for [wave] energy doesn't close before the industry has products ready for prime time."

A further snag in wave energy's great promises, says Feo, is that offshore projects located outside of state maritime boundaries start to wade into murky regulatory waters. The process for obtaining a federal permit is lengthy and convoluted, as two separate regulators—the Federal Energy Regulatory Commission and Minerals Management Service—have been arguing over their joint rights to grant energy development permits in federal waters.

The upside of offshore wave energy, however, is that it gets around the NIMBY issues that can delay or shut down projects, because the giant snake-like conversion structures are typically located so far from shore that they're not visible. The world's first wave power plant, designed by a Scottish firm, got up and running off the coast of Portugal in 2008 and has a production capacity of 2.5 MW of clean energy. The company plans to boost its capabilities to 21 MW. This future capacity combined with other renewable energy resources are projected to save 60,000 tones of carbon dioxide emissions yearly.

Hank Courtright, a division vice-president of the Electric Power Research Institute, said:

> Wave energy is an emerging energy source that may add a viable generation option to the strategic portfolio. The bedrock of a robust electricity system is a diversity of energy sources, and wave energy could provide an energy source that is consistent with our national needs and goals.

Unsurprisingly, the most controversial form of energy used today is nuclear. This low-carbon power source has the potential to provide cheap, virtually pollution-free energy—but it also summons images of mushroom clouds and radioactive fallout. The science behind nuclear technology involves a process called fission or splitting. During this process, energy is extracted from a controlled nuclear reaction involving uranium. This extracted energy is then used to heat water, producing steam to turn turbines that power electric generators. Fission of uranium is currently the only commercial method in use, but others are being explored.

Despite heavy political posturing around nuclear energy's global security risks, which fuels public paranoia, nuclear power is widely in use. This natural resource accounted for 15 percent of the world's electricity by 2005, with the U.S., France and Japan producing over half of worldwide nuclear-generated electricity. The International Atomic Energy Agency reported that by 2007, there were more than 400 nuclear power plants operating in 31 countries in the world.

Canada has 20 nuclear reactors, most of which are located in Ontario, and is the world's largest producer of uranium. Most of the country's aging plants were built in the 1970s and '80s and have unresolved safety issues, including radioactive waste storage, as well as high-costs overruns and underperformance issues.

Yet nuclear is often called "clean" energy because it produces no smog or greenhouse gases linked to global warming, and with new technological upgrades, it holds the promise of being one of the most efficient and least polluting natural energy sources on the planet.

The main problem with nuclear energy is that uranium waste is radioactive and, if stored improperly, could end up in the soil and water system. This waste stays radioactive for thousands of years; if another accident like Chernobyl were to happen, the consequences would again be devastating and long term. Exposure to uranium increases a person's risks of bone and liver cancer as well as blood diseases. Thousands of residents, firefighters and rescue workers were exposed to nuclear radiation following the April 26, 1986, nuclear disaster in Kiev, Ukraine. It's been estimated that 4000 people could eventually die from the exposure.

Current advances in nuclear technology are refining storage safety and recycling techniques, and one day it will be possible to "reclaim" up to 95 percent of nuclear waste. Initial investments are high, but energy costs would go down in the long term.

Western European countries, including Finland, France, Sweden, Spain, Portugal and the

U.K., as well as China and India are all exploring nuclear as an alternative to petroleum. By adding it to their energy mix, they hope to lower energy costs, decrease their fuel dependencies on other nations and meet Kyoto's carbon-reduction mandate.

Meanwhile, after exhausting conventional liquid reserves, petroleum companies in Alberta's oil sands are laboriously preparing to squeeze some of the last and most stubborn billions of barrels of oil from the ground. The job is hard and messy. Oil sands extraction releases about double the greenhouse gas emissions per barrel compared to regular drilling and wells. But with the help of government subsidies, researchers are working on developing a cleanup program that could eventually lower greenhouse emissions by 69 percent per barrel. Current techniques being explored include the production of synthetic gas from coal and other industrial by-products, carbon-capture storage and the use of nuclear reactors to help eliminate harmful emissions released into the atmosphere.

Advances are promising; however, there's strong consensus in the oil industry that the current government penalties for exceeding allowable emissions are too low to drive an industry-wide shift because it's still cheaper to pay their $15 per ton penalty than to install cutting-edge cleanup technologies. The penalty would need to quadruple to provide enough incentive for companies to warrant the steep investment.

In the end, a successful—and equitable—shift to clean energy will require a winning combination of smart government policy-making, tax incentives and subsidies, increased market demand for renewable energy and education of the public and industry stakeholders. In reality, until "green gold" is as shiny and valuable as "black gold," this shift won't likely happen, so the government needs to lead the way.

In the meantime, as cheap oil and gas are depleted, and oil prices skyrocket, the faces of our energy, agriculture, transportation and manufacturing sectors will rapidly change, reverting us back to a more locally driven economy, says Canadian economist Jeff Rubin.

His new book, with a talkshow worthy title, has a compelling message. *Why Your World is About to Get a Whole Lot Smaller: Oil and the End of Globalization* bluntly asks us:

> What do sub-prime mortgages, Atlantic salmon dinners, SUVs and globalization have in common? They all depend on cheap oil. And in a world of dwindling oil supplies and steadily mounting demand around the world, there is no such thing as cheap oil. Oil might be less expensive in the middle of a recession, but it will never be cheap again. Take away cheap oil, and the global economy is getting the shock of its life.

Rubin is optimistic, however, that this shift will take us to a better place—revitalizing local economies and communities, cleaning up our environment and driving healthy economic growth in the renewable energy sectors.

"Our economy will be more diversified and self-sufficient," says Rubin, and higher oil prices will be "better than a hundred Kyotos" in reducing greenhouse gas emissions.

In the future, our priorities in a post-global world will be different. We may buy less and take public transit more, and say "goodbye" to winter strawberries and summer air-conditioning and "hello" to runaway inflation and steep energy premiums, as governments scramble to pay down giant deficits and peak oil plummets and bottoms out. But the world we'll be living in will be far more attractive, sustainable and kind to future generations than the one we're living in now.

As Gaylord Nelson, the co-founder of Earth Day, said, "The ultimate test of a man's conscience may be his willingness to sacrifice something today for future generations whose words of thanks will not be heard."

From where we stand today—morality aside—there may be few other reasonable choices than to cross our fingers, take the lessons that climate change is teaching us and pay them forward, for the sake of our planet—and our children. After all, it's our debt, not theirs, and unless we pay up, the burden will just be passed on. The sooner we roll with these enlightened changes, the better for all. It's not just the smart thing to do. It is our sovereign duty.

As Canadians looking north to lands and waters we want to call our own—lands and waters soon to be slated for major energy development projects and soon to be dramatically transformed by modern climate change—it's incumbent upon us to clean up our act and implore others to tread more lightly. Only then, when national pride and authentic stewardship converge, will we have the right to entitlement. Only then, will we be able to call ourselves truly sovereign.

Conclusion

In the record-breaking summer of 2008, temperatures on Baffin Island hit 27° C, far above the seasonal average of 12° C. Massive flooding shut down a major national park in the Canadian Arctic, following a surge of meltwater that washed out 60 kilometres of hiking trails and destroyed a bridge, leaving 21 tourists stranded and in need of air evacuation.

Auyuittuq National Park, located on southern Baffin Island in the eastern Arctic, spans 19,000 square kilometres of stunning ice-capped vistas, which are slowly disappearing.

Auyuittuq—meaning "land that never melts" in Inuktitut—is a geologic rarity and a natural sanctuary with curving coastlines and fjords, the giant Penny Ice Cap and the highest peaks of the Canadian Shield. Aside from its breathtaking mountain ranges and pristine ice fields, the park also offers visitors who flock there an opportunity to hike the ancient Akshayuk Pass, a traditional corridor used by the Inuit for millennia.

Conclusion

But history is eroding in this primordial place, along with the present-day face of the popular parkland. Scientists point to climate change for the dramatic temperature increases up north, higher than global averages in recent decades, which are melting glaciers and permafrost, spurring flash floods, destroying natural habitats and threatening the future of the north and its people, lands and wildlife.

Jump back a week earlier, when peaking mercury finally tipped the scales of nature's time-honoured balance, causing early snow-cover melt and 20 square kilometres of ice to break off of a Canadian Arctic shelf, triggering this chain reaction. Scientists say it's just the tip of the iceberg.

The great polar thaw of our lifetime is now underway, changing the north—and the world—as we know it, in some ways that are irreversible. It's a one-way trip to a new frontier, but what this destination "hot spot" will look like once we get there is still uncertain.

As circumpolar nations expedite efforts to map the region and assert ownership of oil-rich sea territory—spending billions to build their military, scientific and economic presence in the north—an international storm is brewing over conflicting claims to North Pole waters.

The same year that Baffin Island saw record-breaking temperatures, foreign ministers from the Arctic-5 nations—Denmark, Russia, the U.S., Canada and Norway—gathered for a three-day conference in Ilulissat, Greenland, to hash out an Arctic Nations Pact, which would ostensibly defuse territorial tensions and signal to the world co-operation in the Far North.

At the time, the Danish Foreign Minister Per Stig Moeller told the media:

> We have claims—and the others also have claims. So what I am hoping to get out of this conference is that we agree on the rules of the game. That we do not do anything [that] harms the others until the United Nations has decided who is entitled to what areas of the North Pole.

At the summit's close, on May 28, 2008, the leaders announced the Ilulissat Declaration, which promised "the orderly settlement of any possible overlapping claims" in the northern sea. It also looked toward greater co-operation on maritime security, search-and-rescue missions, environmental protection and patrolling of foreign vessels travelling through the Northwest Passage.

The Declaration stated:

> The Arctic Ocean stands at the threshold of significant changes. Climate change and the melting of ice have a potential impact on vulnerable ecosystems, the livelihoods of local inhabitants and indigenous communities and the potential exploitation of natural resources. By virtue of their sovereignty, sovereign rights and jurisdiction in large areas of the Arctic Ocean, the five coastal states are in a unique position to address these possibilities and challenges.

The declaration also called for the creation of a legal body, such as the Arctic Treaty, under the

Conclusion

umbrella of the United Nations, to monitor commercial shipping, oil and gas activity and scientific ventures in the High North.

Participating delegates applauded the meeting's outcome, but since then, the so-called Arctic-5 have appeared to be locked in a power struggle over who controls the Arctic region and how to divvy up land and water, and the expected windfall of energy resources, as climate change opens the region.

In the summer of 2009, Russia announced its plan to send paratroopers by parachute to the North Pole the following spring, to mark the 60th anniversary of the Cold War achievement of two Soviet scientists. The proposed mission reeks of provocation and is reminiscent of Russia's cheeky submarine flag-planting mission under the North Pole in 2007. Polar pundit Rob Huebert says the brazen act will only increase friction between competing Arctic nations. He told the media at the time:

> The political sensitivity of sending a paratrooper drop at the North Pole? ...[Consider] the political symbolism and the military capability that that shows—it's clear the Russians are very much increasing their assertiveness, and I'd say starting to border on aggressiveness, in terms of their intent to show their ability to have control of the Arctic region.

A record number of foreign ships navigated through Canada's Arctic Archipelago islands, and experts predict a steady growth of traffic through the Northwest Passage and Northern Sea Route, along Russia's Arctic coast, between Europe and Asia, in coming years.

Each voyage through the Northwest Passage weakens Canada's claim that the passage constitutes internal waters, while bolstering the U.S. and Europe's argument that the circumpolar shortcut is an international strait. Canada and the U.S. still "agree to disagree" over who controls this waterway.

By mid-summer of '09, the two countries had announced a joint-mapping survey of the Arctic seabed, slated for August. However, what appeared to be a synergetic move could soon prove to be problematic. One day the two countries could be working together to chart an area of mutual interest; the next day, conflict could arise if overlapping claims are made along the Canada-Alaska border, in the Beaufort sea, which is known to contain rich stocks of extractable oil and gas, experts surmise.

These days, everyone's so busy with their metre sticks and sonic transmitters that they've put their differences on hold to "do science" with one another, says Huebert. But it's only a matter of time before those differences resurface, and when they do, hard science will only go so far to resolving the contentious issues.

"The dispute [over the passage] exists," Foreign Affairs legal expert Caterina Ventura told media, adding that it is "well-managed." She affirmed:

> At one point in time, this is an issue that both Canada and the U.S. will decide to move forward on and find options. But at the moment, we're focusing our efforts on co-operation with regards to mapping of the continental shelf in the area.

Conclusion

That same week, the Canadian government launched a public relations campaign to highlight Canada's independent and on-going efforts to develop and protect its interests in the northern region.

The glossy, patriotic 44-page report—"Canada's Northern Strategy: Our North, Our Heritage, Our Future"—and a new website that tracks the government's latest Arctic policy actions and investments are meant to build on Harper's Arctic agenda, rolled out in 2007 and 2008.

Canwest News Service, wrote:

> [The new report] summarizes the "concrete action" being taken in the four main areas of the government's northern strategy—sovereignty, social and economic development, the environment and governance reform—and casts Canada as a leader on the international stage when it comes to charting the Arctic's future, particularly in an era of rapidly retreating sea ice and rising demand for offshore oil.
>
> Notably, the report downplays the likelihood of international conflict over undersea territorial claims being filed by the five nations with Arctic coastlines...and emphasizes that science and strict protocols under the UN Convention on the Law of the Sea will resolve any disputes about new polar boundaries.

The document asserts, "This process, while lengthy, is not adversarial and it is not a race."

But, clearly it is—and Canada is stepping up its game to pull out from behind and into a more respectable position so that Canadians don't have to trade their spotless snow-capped visions of their "true north" for sunscreen, flood gear, commemorative Inuksuks, border paranoia and a guilty, oil-slicked conscience.

Science and the art of spin can only go so far to truly protect Canada's sovereignty in the Far North. In the end, the debate over northern stewardship will go beyond seabed sediment and ocean law, and will be settled in the political arena, where the international issues of the day are present at the table, says Huebert.

> Are the Americans ratifying conventions or still engaging in dinosaur politics? Are we working out a deal with the Russians and Danes that's bilateral or trilateral? Has the Ukraine put in an application to join NATO? If so, Russia is not going to want to co-operate with that. Has there been another Chernobyl? What China does with Taiwan—that is going to be much more dangerous than whatever the Russians and U.S. might do in the Arctic. The point is, the politics flavouring the environment around us will affect the outcome in the north.

Today's economic crisis is deflecting everyone's attention away from the Arctic, he adds. But every couple of weeks, the issues are reiterated with the latest news of military, scientific or commercial developments in the north. Hopefully, Harper will continue to move his promises forward in spite of the financial crunch we're in. Politics or not, only actions can speak louder than words to get the job done. Huebert concludes:

Conclusion

The negative view here is that they have simply made their promises and think they look good, so they're biding their time until the next election. That's what's happened in the past, but I'm crossing my fingers that's not going to happen now.

Various waves emanating from the Arctic are all coming together. Military might, environmental issues, ocean law, the northern lands, culture and people—they're all interconnected in this sovereignty debate.

That's why the question of "Who's north?" is no small discussion with no easy solution. The question of Canadian sovereignty can't be asked in a bubble, because the answer only exists in the epicentre of a dynamic, shifting world, where climate change, global security, depleting oil, run-away capitalism and endangered people and places shape the reality of our present and possible futures.

To be a truly northern nation will require more than glossy reports, campaign promises and our legal status stamped on paper, because sovereignty is not about impressions, but actions.

In short, Canada can't call itself a northern nation unless it acts like one. Just how our nation will accomplish this remains to be seen. But what will make Canada a truly northern place must go beyond symbolism, history to date and good intentions. True sovereignty only exists in the now. To be authentic, a nation's sovereignty must be lived—in how it listens, debates, reflects and, finally, responds to serve its regions and its people.

Canada must do this in the Arctic, not just for northerners, but for all Canadians and their offspring. Canada's northern sovereignty is not a *fait accompli*; it is an aspiration in the making. It will be determined by how willing—and how well—Canadians protect the north, including its environment and Native people, guard and defend their borders and promote a sustainable world for future generations.

The Arctic can no longer be a ghost arm of the Canadian psyche. The north must be hard-wired into the spirit and body politic of the nation. It must be in our thoughts, our words and our deeds. After all, if the new north is not strong and free, it's not really Canadian, anyway—eh?

Selected References

"Battle for Arctic Heats Up." CBC News, February 27, 2009. Available at http://news.sympatico.msn.cbc.ca/Local/NFLD/ContentPosting?newsitemid=stjohns-f-arctic-sovereignty&feedname=CBC_LOCALNEWS&show=False&number= 0&showbyline=True&subtitle=&detect=&abc=abc&date=True

Berthiaume, Lee. "Obama–Harper: The Issues." In *Embassy Magazine*, February 18, 2009. Available at http://www.embassymag.ca/page/view/obama_harper-2-18-2009

Boswell, Randy. "Canada to Make Groundbreaking Arctic Claim." *National Post,* August 6, 2008. Available at http://www.nationalpost.com/news/canada/story.html?id=705136

———. "Scientists Bolster Claim to Arctic." Canwest News Service, May 19, 2009. Available at http://www2.canada.com/edmontonjournal/news/story.html?id=03ad8ffb-c343-4e67-8284-91b3f765b0b3

"Canada's Claims to Its North." *Canadian Encyclopedia Historica.* Available at http://www.thecanadianencyclopedia.com/index.cfm?PgNm=TCE&Params=A1SEC816206

"Canada's Climate Plan Dishonest." Greenpeace, November 24, 2008. Available at http://www.greenpeace.org/canada/en/recent/canadas-climate-plan-fraud#

Coates, Ken, P. Whitney Lackenbauer, William Morrison and Greg Poelzer. *Arctic Front: Defending Canada in the Far North.* Thomas Allan Publishers, 2008.

Doyle, Alister. "Russia's Seabed Flag Heralds Global Ocean Carve-up." Reuters News Service, August 15, 2007. Available at http://www.minesandcommunities.org/article.php?a=1768

Dufresne, Robert. "Canada's Legal Claims Over Arctic Territory and Water." Library of Parliament, December 6, 2007. Available at http://www.parl.gc.ca/information/library/PRBpubs/prb0739-e.htm

"Energy Future: A Significant Period of Discomfort." Interview with Robert Hirsch, posted on Alianz.com, June 20, 2008. Available at http://knowledge.allianz.com/en/globalissues/safety_security/energy_security/hirsch_peak_oil_production.html

Faris, Stephan. "Forecast: The Consequences of Climate Change from the Amazon to the Arctic, from Darfur to Napa Valley." Henry Holt & Company, 2009.

"Federation of Canadian Municipalities Asks Members to Ban Bottled Water." CBC News Online, March 7, 2009. Available at http://www.cbc.ca/canada/british-columbia/story/2009/03/07/bc-fcm-bottle-water-ban.html

Gore, Al and Davis Guggenheim. *An Inconvenient Truth: Transcript. 2006.* Paramount Home Entertainment. Available at http://www.admc.hct.ac.ae/hd1/blog/gw/An%20Inconvient%20Truth%20Transcript.pdf

Granatstein, J.L. "Does the Northwest Passage Still Matter?" *Globe & Mail*, January 12, 2009. Available at http://www.theglobeandmail.com/servlet/story/RTGAM.20090109.wcoarctic12/BNStory/specialComment/home

Hamilton, Tylor. "Sun Rises on Ontario Solar Farm Industry." *Toronto Star,* April 22, 2008. Available at http://www.thestar.com/comment/columnists/article/416481

Harper, Kenn. "Who's Hans?" In *Canadian Geographic*, 2005. Available at http://www.canadiangeographic.ca/hansIsland/background.asp

Huebert, Rob. "Canada and the Changing International Arctic." In *Northern Exposure: Peoples, Powers and Prospects for Canada's North*. Institute for Research on Public Policy (IRPP), 2008 (pre-released). Available at http://www.irpp.org/books/archive/AOTS4/huebert.pdf

———. "Climate Change and Canadian Sovereignty in the Northwest Passage." In *The Navy League of Canada*, Winter, 2001. Available at http://www.navyleague.ca/eng/ma/papers/huebert_e.pdf

Impacts of a Warming Arctic. Arctic Climate Impact Assessment (ACIA), Cambridge University Press, 2004. Available at http://amap.no/acia/

Kirby, Alex. "Why World's Taps Are Running Dry." BBC News Online, June 20, 2003. Available at http://news.bbc.co.uk/2/hi/science/nature/2943946.stm

Kolbert, Elizabeth. *Field Notes From A Catastrophe: Climate Change—Is Time Running Out?* Bloomsbury Publishing, 2006.

Lorenz, Andrew. "Sovereignty Tussles Over Arctic Territory Threaten to Impede Oil and Gas Exploration" In *Oilweek Magazine*, October 2007. Available at http://www.oilweek.com/articles.asp?ID=471

"Mackenzie Valley Pipeline and Alberta Tar Sands." Sierra Club of Canada, 2009. Available at http://www.sierraclub.ca/national/programs/atmosphere-energy/energy-onslaught/campaign.shtml?x=307

McCarthy, Michael. "Riches in the Arctic: The New Oil Race." *The Independent,* July 25, 2008. Available at http://www.independent.co.uk/environment/nature/riches-in-the-arctic-the-new-oil-race-876816.html